High-Speed Circuit Board
Signal Integrity

For a listing of recent titles in the *Artech House Microwave Library,*
turn to the back of this book.

High-Speed Circuit Board Signal Integrity

Stephen C. Thierauf

Artech House, Inc.
Boston • London
www.artechhouse.com

Library of Congress Cataloguing-in-Publication Data

A catalog record for this book is available from the U.S. Library of Congress.

British Library Cataloguing in Publication Data

A catalog record for this book is available from the British Library.

Cover design by Igor Valdman

International Standard Book Number: 1-58053-131-8

10 9 8 7 6 5 4 3 2 1

To Ann, Christopher, and Kevin

Contents

Preface

This is a book for engineers designing high-speed circuit boards. To the signal integrity engineer, this book will be a handbook of formulas and terminology as well as a refresher of basic electrostatic and electromagnetic principals. The high-speed circuit designer will find this book an easy entry into the electromagnetics and physics of high-speed signaling. It introduces concepts fundamental to high-speed signaling, such as lossy transmission line behavior, skin effect, and the characteristics of laminates and surface mount capacitors. The focus throughout is on the effects of dielectric and conductor loss on signal quality, with a particular emphasis on serial differential signaling. Coupling between transmission lines (especially in the context of crosstalk and odd/even modes) is discussed. Besides being useful in serial signaling, this has application to multiconductor busses.

Reflections on transmission lines are only superficially covered in this text. This topic has been extensively covered in the literature, and the reader of this book is assumed to be familiar with the creation and mitigation of reflections on transmission lines. However, the proper routing and termination of differential pairs has not been as well covered in the literature and so is discussed in Chapter 8.

Similarly, power supply decoupling has been thoroughly discussed elsewhere, so the coverage in this book is brief. Instead, the focus here is on managing return paths (something often not well covered) and the electrical characteristics and behavior of capacitors. The material in Chapter 10 is a concise catalog of essential electrical characteristics of discrete capacitors, with a focus on surface mount technology.

The fundamentals of resistance, capacitance, inductance, and loss calculations presented in Chapters 2–5 are illustrated with practical worked examples that may be used as templates to solve similar problems.

Many simple formulas are presented to allow hand calculation of resistance, capacitance, inductance, and impedance. These types of calculations are helpful in developing intuition and in approximating beforehand the results to be expected from field solving software, circuit simulation tools, and laboratory measurements.

Extensive references are given at the end of each chapter, providing the interested reader the opportunity to dig deeper. The references intentionally span classic, older works (some of which were written in the 1950s, but most of the older ones are from the 1960s and 1970s) as well as modern works. The older references are valuable, as they are the original works often cited by others, sometimes without the proper context. Although long out of print, the selected older references are generally available secondhand and are worthy additions to the signal integrity engineer's library. Particular mention is made of Johnson's *Transmission Lines and Networks* (published in 1950, referenced in Chapter 2) and Skilling's *Electrical Engineering Circuits* (1965, see the references in Chapter 3) and *Transient Electric Currents* (1952). These books are succinct and in my view remain unmatched. Miner's *Lines and Electric Fields for Engineers* (1996, first referenced in Chapter 3) is the one electromagnetics textbook every signal integrity engineer should have in his or her library.

I'm indebted to my friends and coworkers for their support, encouragement, and help during the creation of this book. Special mention must be made of the assistance, perspective, and advice provided by my colleagues Jeff Cooper, Ernie Grella, and Tim Haynes. Special thanks also goes to Fahrudin Alagic for his many months of precise laboratory measurements that support the material appearing in Chapters 5–7. I'm grateful to all of those who suffered through early versions of the manuscript for their constructive remarks. I'm also obliged to the anonymous reviewer for his insightful comments. All of these comments were most helpful and have resulted in an improved text. Of course, any inaccuracies or errors that made it into the text are my doing and in no way reflect on the reviewers.

Finally, I'm especially grateful to my wife Ann for her understanding, patience, encouragement, and unflagging support throughout the many long hours it took to create this work. This book would not have been possible without her.

CHAPTER 1

Characteristics and Construction of Printed Wiring Boards

1.1 Introduction

This is a book about high-speed signaling on printed wiring boards (PWBs). The physical construction of PWBs determines the conductor's resistance (discussed in Chapter 2), its self capacitance (covered in Chapter 3) and inductance (Chapter 4), and the coupling to neighboring conductors (Chapters 5 and 9). At the high frequencies of interest in this book, these electrical primitives appear on a PWB as distributed rather than lumped elements, giving rise to transmission line behavior.

It is thus necessary for the high-speed circuit designer to have an understanding of how PWBs are constructed and a sense of the trade-offs fabricators must make when manufacturing high-density, high-layer count PWBs. This chapter summarizes those characteristics impacting the electrical characteristics of PWBs and introduces some of the terminology used in the PWB design industry.

The larger PWB fabricators provide *design for manufacturability* (DFM) documents (see [1, 2] to cite just two examples) that detail the dimensional and many of the practical requirements necessary to create PWB artwork for their facility. These documents are helpful in understanding the practical state of the art in such things as via size, layer count, and trace width and spacing and can act as a primer to those unfamiliar with PWB technology. Additional underlying detail that is somewhat general in nature may be found in [3, 4].

1.2 Unit System

The PWB industry nearly universally uses an inched-based measuring system rather than the metric system. Trace width and length and dielectric thickness are thus specified in decimal fractions of an inch, as are most component dimensions. However, many micopackage dimensions (most notably the pin or ball pitch) are specified with metric millimeters, and the trace thickness is specified in ounces (relating to the amount of copper plating, as described in Section 1.4). The Appendix tabulates some common conversion factors, but here it's noted that $1m = 39.37008$ in and 1 mil = 0.001 in. Therefore, 1 mil = $0.002539 \sim 0.00254$ cm = $0.02539 \sim 0.0254$ mm.

Example 1.1

A ball grid array micropackage (BGA) has solder balls on a 1-mm pitch. What is the pitch in mils?

Solution

Referring to Appendix A, to convert from inches to millimeters, the value in inches is multiplied by 25.4. The 1-mm ball pitch therefore is equivalent to: $\dfrac{1 \text{ mm}}{25.4 \text{ mm/in}} =$ 39.37×10^{-3} in $= 39.37$ mils. As there are not precisely 39.37 in per meter, the conversion factor is not precisely 25.4 mm/in. This error is often inconsequential but can be important over large distances.

1.3   PWB Construction

The typical multilayer PWB is formed as a stack of alternating layers of prepreg mats and laminate sheets. The general idea is shown in Figure 1.1.

The prepreg mats are a weave of glass fiber yarns preimpregnated (hence *prepreg*) with a resin that is intentionally allowed to only partially cure. The sheets come in many stock sizes and yarn *styles* (classified by the number and diameter of the glass threads, the weave, and the percentage of resin impregnation) and serve to strengthen the resin. The typical resin content of the mats is in the 45% to 65% range.

Copper foil is attached to one or both sides of fully cured prepreg sheets to form the laminate sheets (also called *cores*). Similar to the prepreg mats, cores come in standard stock sizes and thicknesses, from which the fabricator must choose to construct a PWB. It's common for outer layers (such as layers L1 and L6 in Figure 1.1) to be formed on prepreg [1], but some manufacturers prefer to form the outer layers on cores.

To form the composite PWB structure, a stackup of prepreg mats and laminate cores are heated under pressure. This causes the partially cured prepreg to flow and bond to the cores. The prepreg cures are cooled, thereby forming the completed PWB structure.

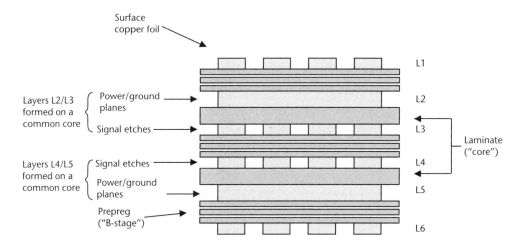

Figure 1.1   Multilayer PWB stackup.

1.3.1   Resins

Several resin systems are used to form prepreg and laminate sheets, with the FR4 epoxy resin system probably being the most popular.

The generic specification *FR4* refers to a specific fire-retardant level rather than to a specific laminate chemistry. The term *standard FR4* is a misnomer: Fabricators have many laminate systems from which to choose that meet FR4 fire specifications, each with unique electrical and mechanical characteristics. Fabricators select the laminate system based on assembly issues (especially temperature extremes) and their history with a particular laminate for a given set of electrical requirements.

The FR4 epoxy can be blended with tetrafunctional or multifunctional resins [5, 6] to improve the material's mechanical characteristics, most notably the coefficient of expansion, the *glass transition temperature*, and the rate of moisture absorption [5].

The coefficient of expansion in height—that is, the Z axis, *CTE(Z)*—is an indication of how much the board will enlarge and contract in thickness with temperature changes. Increasing the resin's temperature above its glass transition temperature (Tg) causes the resin to change from its crystalline state to a more fluid, glassy state. Additionally, CTE(Z) is not constant across temperature: increasing temperature above Tg results in a rapid increase in CTE(Z) [7].

Fabricators will recommend the use of a *high Tg* laminate (those with Tg greater than about 180°C [5]) if the PWB will be exposed to high temperatures during assembly or rework operations. This is important because thermally induced stresses can lead to plated through hole failures (see Section 1.5), especially in thick boards having a large CTE(Z).

The drilling operation is another consideration for the fabricator when choosing a laminate. The typical FR4 resin system is relatively easy to drill. High Tg laminates tend to be harder and more brittle, making drilling more difficult. On the other hand, high-speed drilling of vias may warrant the use of a high Tg material to avoid *drill smear*. This is a result of a softening of the laminate near the hole due to drilling induced local heating that momentarily raises the temperature above Tg (Tg ~ 130°C for FR4 [5]).

1.3.2   Alternate Resin Systems

Resins other than the standard FR4 epoxy types are available to form PWB laminates and prepregs. These alternate systems have higher Tg than the tetrafunctional or multifunctional FR4 resins, and superior electrical characteristics. Of principal interest are the values for the dielectric constant (ε_r discussed in Chapter 3) and loss tangent (a parameter describing the amount of energy lost in the dielectric at a specific frequency, also discussed in Chapter 3).

Some of the alternate resin systems include GETEK® (a registered trademark of General Electric Company), MEGTRON® (a registered trademark of Matsushita Electronic Materials), BT (a blend of bismaleimide/triazine such as Allied Signals G200 [8]), polyamide, and cyanate ester resins. As with the FR4 epoxies, these resins are reinforced with glass or other fibers.

GETEK and MEGTRON are blends of polyphenylene oxide and high Tg epoxy reinforced with E glass [9, 10]. These laminates have lower ε_r and loss tangent values than FR4 systems and a lower CTE(Z).

A blend of bismaleimide and cyanate ester resins commonly called BT was originally introduced by Mitsubishi Gas and Chemical Company but is now available from several laminate vendors. This resin system has electrical characteristics somewhat superior to FR4 and is used extensively in the micropackaging industry due to its higher Tg and superior moisture absorption qualities.

Polyamide resins generally have lower ε_r and loss tangent values than FR4 resins and have a significantly higher Tg. These desirable characteristics are somewhat offset by polyamide's affinity for moisture. The ability of these laminates to withstand high temperatures suits them to aerospace applications and commercial test equipment, such as burn-in chamber circuit boards where semiconductors are life tested at high temperatures. Polyamide is also extensively used in the flexible circuit board industry.

Cyanate ester resins have superior electrical characteristics to polyamide and exhibit lower moisture uptake. They are often used in RF applications, but this material is not as suited as other materials to form multilayer stackups [11], so they are not as popular in high-speed digital design work.

The RO4000® series laminates from Rogers Corporation are reinforced hydrocarbon/ceramic materials that are finding increasing use in high-speed digital signaling. These materials have a very high Tg, low loss tangent, and a stable ε_r up to at least 10 GHz [12].

A synopsis of these resin systems in laminate form as represented by Nelco Park [13], Isola-USA [14], Matsushita, and Rogers Corporation appears in Table 1.1. A more complete listing showing various laminate systems from several vendors appears in Chapter 3.

As shown, the multifunctional FR4 epoxies have the lowest Tg and highest ε_r/loss tangent values. Of the resins, polyamide has the highest Tg and ε_r/loss tangent values, second to the cyanate ester resin system. The Rogers RO4350® has the highest Tg and lowest ε_r and loss tangent value.

Table 1.1   Alternate Laminate Systems

Trade Name	Chemistry	Tg	ε_r/loss tan	ε_r/loss tan	Vendor
		C°	1 MHz	1 GHz	
N7000-1	Polyamide	260	4.3/0.013	3.7/0.007	Nelco
P97	Polyamide	260	4.4/0.014	4.2/0.014	Isola
N8000	Cyanate Ester	250	3.8/0.008	3.5/0.006	Nelco
N5000	BT	185	4.1/0.013	3.8/0.010	Nelco
G200	BT	185	4.1/0.013	3.9/0.009	Isola
N4000-6	Multifunctional	180	4.4/0.023	3.9/0.012	Nelco
Megtron	PPO/Hi Tg Epoxy	180	3.8/0.010	3.75/0.011	Matsushita
FR404	Multifunctional	150	4.6/0.025	4.25/0.014	Isola
RO4350	Ceramic	>280		3.48/0.004 (10GHz)	Rogers

1.3.3 Reinforcements

Fibers (usually from a form of glass) are used to strengthen the resins, but adding them changes the electrical and mechanical characteristics of the composite structures roughly in proportion to the amount of fiber to resin (the *glass-to-resin ratio*). As shown in Table 1.2, the glass fibers have a higher ε_r but superior loss tangent values than the resins.

High glass content improves the composite's CTE(Z), thereby helping to prevent via cracking during high-temperature assembly and rework operations. However, a high glass content increases ε_r and lowers the loss tangent [15, 21]. Generally, higher ε_r is a disadvantage in high-speed PWBs, as that increases capacitive coupling between conductors and tends to result in thicker stackups for a given impedance. Alternatively, for a given stackup thickness, higher ε_r results in narrower trace widths, thereby increasing conductor loss (described in Chapters 2 and 5). Lower loss factors are advantageous, as they improve high-frequency signal qualities (as described in Chapters 3, 5, and 7).

The relationship between resin content and ε_r is generalized in Figure 1.2 for FR4.

The prepregs aggregate ε_r value approaches that of the just the resin for low glass-to-resin ratios and approaches that of the glass itself as more glass is added and the glass-to-resin ratio increases. Figure 1.2 points out the difficulty in judging between laminates simply by comparing published ε_r values, as some manufacturers specify a "worst case" ε_r (i.e., low resin content) while others publish an ε_r corresponding to a higher resin content value (often 50%).

Various glass fiber types are available to reinforce the resin. The most common is *E* glass (electrical grade), which is commonly used throughout the plastics industry. This glass fiber was specifically designed for electrical use, but its versatility has made it suitable for reinforcing a range of plastics. This broad adoption beyond use in the PWB industry is responsible for the low cost of E glass [4]. It primarily consists of silicon oxide, aluminum oxide, and calcium oxide.

Table 1.2 Resin and Reinforcement Properties at 1 MHz

Material	ε_r	Loss	CTE(Z) Parts Per Million (PPM)/C°	Moisture Absorption (%)
E Glass	6.2	0.004	5.5	
S Glass	5.2	0.003	2.6	
Thermount®	3.9	0.015	−4.5	0.44
FR4 epoxy resin	3.6	0.032	85	0.7
BT resin	3.1	0.003		
Polyamide resin	3.2	0.02	50	0.9
Cyanate Ester resin	2.8	0.002	50	0.5

Source: [4, 7, 15–20].

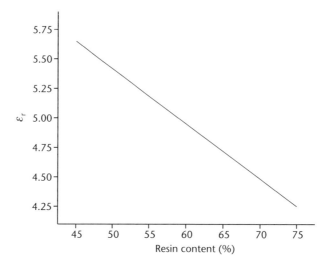

Figure 1.2 Relationship between resin content and ε_r for FR4.

A second glass fiber finding increasing use in PWBs is *S* glass (structural grade). This glass fiber was specifically developed for high-strength reinforcement applications and also consists of silicon oxide and aluminum oxide, but it uses magnesium oxide in place of the calcium oxide [6]. It's stronger than E glass and has a lower ε_r, but it's not as widely used and so is more expensive (about four times that of E glass [6]).

A nonwoven Aramid fiber called Thermount [17] is finding increasing use as a reinforcement to epoxy, polyamide, cyanate ester, or Teflon® resins in PWB applications. Thermount is comprised of very short *Kevlar*® fibers [5]. It offers lower ε_r than either E or S glass and has a negative CTE(Z), which can be advantageous in reducing the expansion of the composite structure [4]. Thermount, Kevlar, and Teflon are registered trademarks of E. I. Dupont de Nemours & Co., Inc.

1.3.4 Variability in Building Stackups

In producing a multilayer PWB as depicted in Figure 1.1, the fabricator must decide on the thickness of the laminate and the styles, thickness, and the number of prepreg mats to use to form each layer. The laminate sheets tend to have lower resin content than the prepreg, so the laminates usually have higher ε_r than the prepreg sheets. The way in which the fabricator chooses to form the stackup is fundamental in that it determines ε_r and the loss tangent for a particular layer. One fabricator may choose to use several thin, high-resin-content mats resulting in a lower overall ε_r, while another prefers to use a single, thicker mat having lower resin content which will yield a higher ε_r to get the same overall thickness.

The vendor's latitude in making all of these choices means that ostensibly identical PWBs fabricated by different vendors will quite naturally have different electrical properties. These trade-offs are discussed in Chapter 9.

1.3.5   Mixing Laminate Types

It's not necessary for the cores to all be the same laminate material. Historically this hybrid-type construction has been expensive and not widely used in the commercial digital PWB industry, but in recent years it's become somewhat more mainstream. In spite of its fabrication complexity, it can be cost effective to use a lower cost, lower performance laminate throughout most of the stackup and strategically mix in a few expensive, higher performance layers only where needed. This is especially attractive for stackups that have many layers, where only a few carry high-frequency, loss-sensitive signals. In these situations, the complexity of fabricating a stackup containing different materials costs less than making the high layer count stackup entirely from the high-performance, expensive laminate. The materials chosen must have similar CTE values [22, 23] so not all laminate types can be mixed.

Also note that the copper thickness need not be the same throughout the stackup. Having different copper thickness on various layers is common in situations where the power/ground planes must be thick for proper power supply distribution but the signal traces need not be. As described in Chapter 2, at high frequency the skin effect causes signal currents to migrate to the conductor's surface. This means thick traces do not necessarily have a loss advantage over thinner ones. An additional advantage of using thin copper for signal traces is that it's easier to retain a truly rectangular shape when etching the thinner copper. This has loss, coupling, and impedance advantages (see Chapter 9). Using thinner copper for the signal traces can help reduce the stackup's thickness, but using thin copper for the power/ground planes reduces their ability to wick heat from the pins of an integrated circuit or field effect transistor (FET) (as is found in switching power supplies or when FETS are used in power supply sequencing circuits).

To avoid warpage, manufacturers favor *balanced* stackups, where the thin and thick layers are distributed symmetrically about the stackup's center, but this also applies to the distribution of laminate types.

1.4   PWB Traces

Copper traces are used to form the PWB conductors, either of the board's surface (*microstrip* or *embedded microstrip*) or buried within the PWB as *stripline* (see Figure 1.3).

To properly model high-frequency conductor losses, it's important for the high-speed circuit designer to understand the process used to form a trace. The

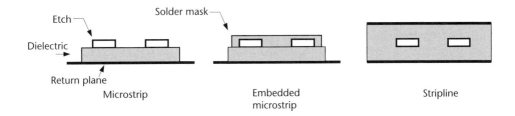

Figure 1.3   Microstrip and stripline defined.

nature of the multilayer PWB fabrication process is such that the mechanical characteristics of the inner layer copper is different from the copper on the board's outer surfaces. The outer layers are plated, while the inner ones are not. Copper cladding is discussed in this section. Plating and the consequences to outer layer conductors are discussed in Sections 1.4.2 and 1.6.

1.4.1 Copper Cladding

The copper cladding attached to laminate sheets is created by either an electrodeposition or rolling process [24, 25]. These processes create copper foils with different surface roughness. As is shown in Chapter 2, accounting for surface roughness is important when computing skin effect losses.

The electrodeposited process (ED) creates copper foil by a plating process that forms a copper sheet by extracting copper from solution onto a rotating drum [26]. The foil side in contact with the drum is smoother than the other surface. In contrast, the rolling process starts with a copper ingot that is passed through rollers multiple times until it is reduced to the desired thickness. This process creates foil equally smooth on both sides, and smoother than that of electrodeposited copper. A smooth surface is advantageous when signaling at high frequency because (as explained in Chapter 2) the ac resistance will be lower with a smooth surface than it will be with a rough one. This makes rolled copper trace electrically preferred over ED at high frequencies. However, the greater coarseness of ED foil allows the copper to better adhere to the substrate, giving ED foils higher *peel strengths*. Foils with higher peel strengths have better adhesion and so are less likely to lift off from the laminate during soldering or rework operations.

To promote adhesion with the laminate material, both types of foils are roughened on one side (or sometimes both sides) to increase surface area. There are many techniques available for fabricators and laminators to use [27], each producing different copper grain sizes and shapes. Surface roughness is measured as the root-mean-square (RMS) height of the irregularity above the surface.

As shown in Table 1.3, in general, even after processing, rolled copper has a lower surface roughness than ED.

The data in Table 1.3 should only be taken as representative. Actual values depend on processing and will vary between manufacturers.

Because the CTE of copper foil is actually lower than that of the laminate, thermally induced stresses can cause the connection to a via to fracture over time or with repeated thermal cycling. High-temperature elongation (HTE) foils can be used to mitigate this susceptibility to stress [28]. These foils are also sometimes called *class 3 foils* after the Institute for Interconnecting and Packaging Electronic Circuits (IPC) industry standards group designation [29]. Foils in this category have a higher CTE

Table 1.3 Typical Copper Foil Characteristics

	Average Thickness (mils)	ED μ-inches (RMS)	Rolled (Treated Side) μ-inches (RMS)
Half ounce	0.65	75–100	50–60
One ounce	1.4	95	50–60

Source: [24, 25].

than the *class 1 foils* that are in general use and more closely match the laminate's CTE. The use of HTE foils is becoming common, especially on higher performance resin-based laminates, but they are generally not used with laminate systems having a low CTE(Z), such as the Rogers 4000® series materials.

1.4.2   Copper Weights and Thickness

The thickness of the copper foil is usually specified by its nominal weight in ounces per square foot of area. Table 1.4 shows the relationship of weight to nominal and minimum thickness as specified by the IPC [21].

Notice that due to plating, the external conductors (i.e., microstrip) will usually be thicker than the inner layers (stripline) of the same weight.

1.4.3   Plating the Surface Traces

A plating process usually forms surface traces where copper is selectively plated on top of the thin foil present on the PWB's surface. The traces thus formed are protected from the subsequent etching step by coating the traces with either a metal (tin or tin/lead) or a nonconducting photoresist [4]. This is visible in Figure 1.4, which shows the copper trace with a plating material on top of the base copper.

1.4.4   Trace Etch Shape Effects

The shape of the trace is a factor in determining its impedance and resistance, and nearly all hand formulas for computing impedance assume a rectangular trace. Rectangular trace shapes are also usually assumed when field-solving software is used to calculate impedance. However, the etching process attacks the copper both vertically and horizontally, resulting in traces that are roughly trapezoidal in shape. This vertical *over etching* is numerically described by the *etch factor,* which is the ratio of the conductor's thickness to the amount of copper that has been undercut:

$$EF = \frac{t}{uc} \qquad (1.1)$$

Referring back to Figure 1.4, t is the trace thickness and uc is the amount the copper trace is undercut on one side [1].

Table 1.4 Relationship Between Copper Weight Specification and Thickness

Weight Specifier (oz)	Nominal Thickness (mils)	Minimum Internal Layer Thickness (mils)	Minimum External Layer Thickness (mils)
$\frac{1}{4}$	0.35	0.25	0.8
$\frac{1}{2}$	0.70	0.50	1.30
1	1.4	1.0	1.8
2	2.8	2.2	3.0

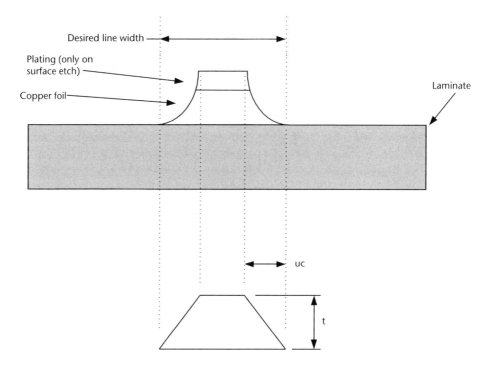

Figure 1.4   Typically shaped trace illustrating the etch factor.

The trace's final shape is dependent on processing and the thickness of the metal being etched. The trace shape and the amount of resin forced between traces during lamination will vary between manufacturers and the PWB layout. Thick, narrow, fine-pitched traces are more likely to be over etched than are thin traces on a wide pitch. For this reason, it's best to use half-ounce or thinner copper when specifying a narrow (~ 5 mils wide) controlled impedance trace [1, 30].

Overetched lines (i.e., those having low etch factors) will have increased line resistance and inductance, but lower capacitance, than expected. The net result is higher impedance [31, 32] and conductor losses than predicted by hand calculations or by field-solving software relative to rectangular shapes. However, the choice of copper type (ED or rolled) is a larger contributor to losses than the trace shape [33].

1.5   Vias

Vias are used to connect traces appearing on different layers. A stackup with vias connecting L1 to L3 and another connecting L3 to L4 is shown in Figure 1.5.

Vias are formed by drilling (either mechanically or with a laser) a hole partially (*blind vias*) or completely through the PWB stackup (*through hole vias*). Those vias that are to be made conductive are then plated. Conductive holes passing entirely through the PWB are called *plated through holes* (PTH).

An annular ring of copper (called a *land* or *pad*) surrounds the via to insure that even with some layer-to-layer misalignment, the drilled hole can still contact the trace on each of the required layers. An off-center via is still able to make complete electrical contact with the copper trace, as shown in Figure 1.6 [1].

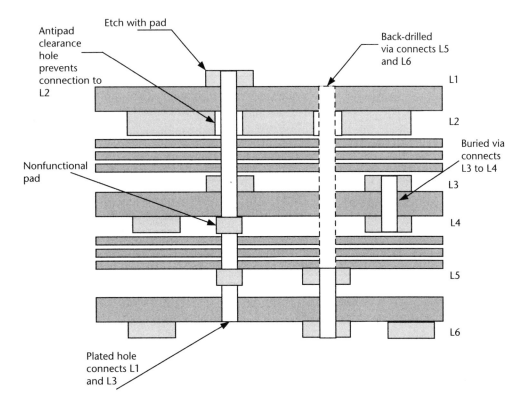

Figure 1.5   Vias connecting L1 to L3 and L3 to L4.

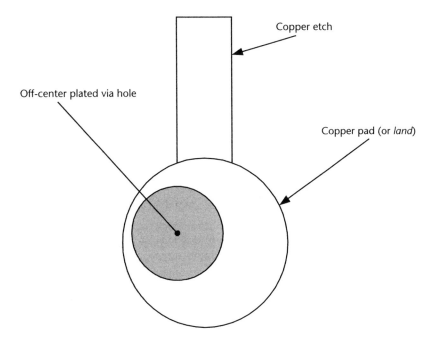

Figure 1.6   Pad allows misaligned via to make contact with signal trace.

These pads are usually only placed on the layers where the via is to connect to a trace, but they can also be placed on the via at where the via does not connect to a trace. These *nonfunctional pads* are visible in Figure 1.5 and serve to anchor the via in the stackup. They are more common on thick boards because vertical stresses caused by thermal expansion are generally greater there than on thin PWB having only a few layers. As discussed in Chapter 6, these nonfunctional pads are often electrically benign, but they do increase a via's self capacitance, which can be detrimental in low-jitter, high-frequency systems.

A clearance hole (sometimes called an *antipad*) is made in the power and ground planes when it's necessary for a via to pass though the plane without making contact. As shown in Chapter 6, the size of this antipad is an important factor in determining the amount of capacitive coupling that occurs from the via to the plane. This is an important consideration in high-speed interconnect. The antipad construction appears in Figure 1.7.

A *thermal relief* pad (diagrammed in Figure 1.8) is placed on the plane in those situations when the via is to connect to the plane. This helps to thermally isolate the via from the plane, preventing the plane from acting as a heat sink and wicking away heat from the via during soldering.

Notice that the connection is made by four tabs from the via body to the plane, increasing the via's resistance and inductance.

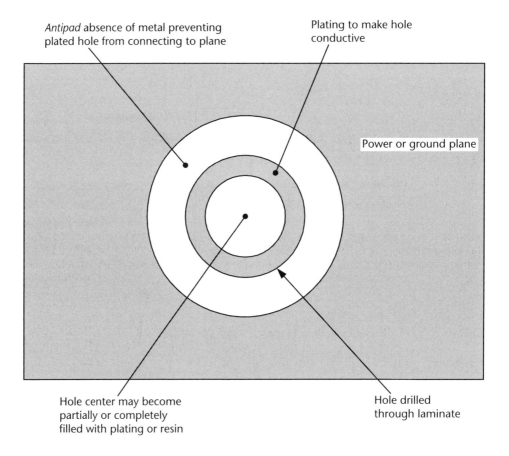

Figure 1.7 Antipad construction.

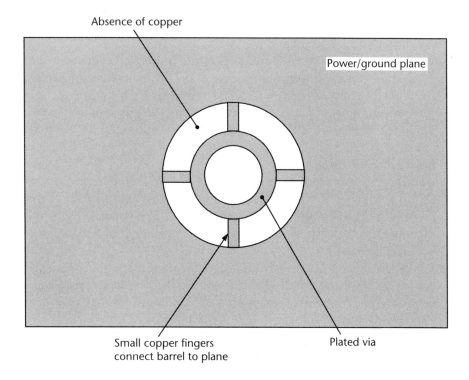

Absence of copper

Power/ground plane

Small copper fingers
connect barrel to plane

Plated via

Figure 1.8 Thermal relief via.

1.5.1 Via Aspect Ratio

Vias that are too long relative to their diameter make the PWB difficult to manufacture and can pose long-term reliability problems.

The ratio of the PWB thickness to a via's outside diameter is a commonly used metric called the *via aspect ratio* (or sometimes simply the *aspect ratio*), (1.2):

$$Aspect\ Ratio = \frac{PWB\ thickness}{Via\ O.D.} \tag{1.2}$$

Fabricators use the via aspect ratio as one way to categorize a PWB's complexity to be manufactured. Each fabricator will set their own aspect ratio limits based on their technical prowess and historical ability. It's generally easy to find fabricators willing to manufacture low-aspect ratio boards, but as the aspect ratio increases the cost per board goes up and the number of capable shops decreases. Low-aspect ratio boards are thus easy to second source and will cost less than those with high-aspect ratios.

Low-aspect ratios (currently considered to be under about eight) are generally the easiest to produce; at the moment, aspect ratios from eight to roughly 10 are considered mainstream and ratios greater than 10 are presently considered to be in the difficult or advanced category. At a significantly higher cost, some advanced fabricators can reliably manufacture boards with aspect ratios approaching 14. This is only a guide; the categorization varies between manufactures and is somewhat dependent on laminate type (especially for the highest aspect ratios).

1.6   Surface Finishes and Solder Mask

Traces appearing on the boards' surface are protected from corrosion by the application of a *surface finish*. Elimination of corrosion aids in soldering and improves the reliability of soldered connections. Many surface finishes are available [34], and the correct choice for a particular situation is determined by such factors as the amount and type of rework the board will be subjected to, the degree of surface coplanarity required, and the shelf life (time the fabricated bare boards can wait before they are assembled) [35].

Some of the more common surface finishes include hot air solder leveling (HASL), which is a coating of solder that can vary significantly in thickness across the board; organic solderability preservative (OSP), which is a very thin application of an organic material; electroless nickel immersion gold (ENIG); or immersion silver or immersion tin.

One would expect that the choice of surface finish would impact high-frequency conductor losses, and indeed this is so. The highest losses have been reported (in descending order) with immersion gold, followed by immersion tin, OSP, and immersion silver having the lowest losses [33].

Solder mask is another coating that is applied to a board's surface. This is sometimes confused with surface finish, but the two are quite distinct. Where surface finish coats all exposed copper to prevent corrosion, solder mask is a thin coating of epoxy placed everywhere on the board except where electrical connections are to be made (such as component solder pads and gold-plated fingers for edge connectors).

Solder masks electrical characteristics are described in Chapter 9, but here it's noted that solder mask comes in several varieties [36, 37] and can increase capacitive coupling between surface traces and cause an increase in loss.

1.7   Summary

Many processes and variations on processes are in use throughout the PWB fabrication industry. This makes it difficult to discuss PWB fabrication characteristics in all but the most general way, yet these details affect the PWBs high- frequency electrical characteristics. The high-speed circuit designer must understand the way that copper is etched and cleaned and the way in which the fabricator has chosen to create the stackup to insure proper high-frequency modeling.

A seeming plethora of laminate systems are available that cover a wide range of cost and performance. This includes the well-known FR4 system and variants on this chemistry incorporating multifunctional and tetrafunctional epoxies. Other systems are available using cyannate ester and polyamide resins, among many others.

To add strength, the resins are usually reinforced with a glass fiber mat of E or S glass, but other reinforcements are sometimes used. The glass fibers have higher ε_r than the resins, and the mat's glass-to-resin ratio determines the ε_r and loss tangent value. A mat consisting mainly of resin will have a low glass-to-resin ratio and thus an ε_r value more like the resin than the glass.

Generally the higher performance laminates cost more and have higher fabrication costs than FR4. The fabrication costs of the higher performance laminates can drive up the delivered bare board cost to many times that of a comparable FR4 board. These costs will vary between manufactures based on their experience and history with a given laminate system. When bringing on second fabrication sources, it's sometimes more economical to specify critical electrical parameters (such as line width, loss, impedance, and time of flight) rather than to specify a specific laminate type. This will allow each fabricator the latitude to select between electrically equivalent laminate systems that yield best for them. Conversely, specifying a particular laminate system in detail (including the prepreg type) will yield better matched boards between vendors but may not be as economical or be very welcomed by a second source vendor.

Etching, plating, and surface treatment factors must be carefully considered when building loss models so as to properly account for skin effect losses.

References

[1]   Merix Corp., *Design for Manufacturability of Rigid Multi-Layer Boards*, Revision 7/99, Forest Grove, OR, July 1999.

[2]   Sanmina-SCI, "Printed Circuit Board Design for Manufacturability Guidelines (Document PCB-PED-07.8.9)," http://www.Sanima-SCI.com.

[3]   Coombs, Clyde F., *Printed Circuit Hand Book*, 5th Ed., New York: McGraw Hill, 2001.

[4]   Jawitz, Martin W., *Printed Circuit Board Materials Hand Book*, New York: McGraw Hill, 1997.

[5]   Ehrler, S., "A Review of Epoxy Materials and Reinforcements," *EIPC Summer Conference*, Copenhagen, Denmark, June 2001. Republished in *PC FAB*, April (pp. 32–38) and May 2002 (pp. 32–36).

[6]   Jorgenson, C., "Is FR4 Running Out of Gas?" *Printed Circuit Design*, September 2000, p. 10.

[7]   Seraphm, D. P., et al., "Printed-Circuit Board Packaging," in *Microelectronics Packaging Handbook*, R. Tummala and E. Rymaszewski (eds.), New York: Van Nostrand Reinhold, 1989, pp. 853–921.

[8]   "G200 BT/Epoxy Laminate and Prepreg," Data Sheet No. 5027/2/99, Allied Signal Laminate Systems, 1999.

[9]   "Megtron(r) PPO/Epoxy Resin System," Data Sheet No. MEM-DS-004 Rev 08, Matsushita Electronic Materials, Inc., January 30, 2003.

[10] "Epoxy/Polyphenylene Oxide Resin," Data Sheet Revision R, GE Electro Materials, November 21, 2001.

[11] Merix Corp., *Applying High-Frequency Materials in Wireless and other RF Applications: Materials and Bonding Agents*, applications note, Forest Grove, OR.

[12] Rogers Corp., "RO4000 Series High Frequency Circuit Material Data Sheet," No. 92-004, Advanced Circuit Materials, Chandler, AZ.

[13] Nelco Products, Fullerton CA.

[14] "FR408 Epoxy Laminates and Prepreg," Data Sheet No. 5035/3/01 Isola-USA, La Crosse, WI.

[15] Mumby, S. J., "An Overview of Laminate Materials with Enhanced Dielectric Properties," *Journal of Electronic Materials*, Vol. 18, No. 2, 1989, pp. 241–250.

[16] Mumby, S. J., "Dielectric Properties of FR-4 Laminates as a Function of Thickness and the Electrical Frequency of Measurement," paper IPC-TP-749, *IPC Fall Meeting*, Anaheim, CA, October 24–28, 1988.

[17] Khan, S., "Comparison of the Dielectric Constant and Dissipation Factors of Non-Woven Aramid/FR4 and Glass/FR4 Laminates," Technical Note, Dupont Advanced Fibers Systems Division, Richmond, VA, Sept. 1999.

[18] Shugg, W. T., *Handbook of Electrical and Electronic Insulating Materials*, New York: IEEE Press, 1995.

[19] "E, R, D Glass Properties," Technical Data Sheet, Saint-Gobain/Vetrotex Corp., March 2002.

[20] Barker-Jarvis, James, et al., "Dielectric and Magnetic Properties of Printed Wiring Boards and Other Substrate Materials," NIST Technical Notes 1512, U.S. Department of Commerce, Bolder, CO, March 1999.

[21] Institute for Interconnecting and Packaging, *Electronic Circuits Generic Standard on Printed Board Design, IPC-2221 2215*, Northbrook, IL, February 1998, http://www.ipc.org.

[22] Merix Corp., "Design Advantages of Using High PerformanceMaterials," Applications Note, Forest Grove, OR.

[23] Weis, V., "Combining Dielectrics in Multilayer Microwave Boards," Application Note, Arlon Materials for Electronics, Providence, RI, October 22, 1997, http://www.arlon-med.com/aboutus.html.

[24] Merix Corp., "Electrodeposited vs. Rolled Copper," Applications Note, Forest Grove, OR, 1997, http://www.merix.com.

[25] Rogers Corp., "Copper Foils for Microwave Circuits," Applications Note No. 92-243, February 2000, http://www.rogers-corp.com.

[26] GE Electromaterials, "The Manufacture of Laminates," Technical Paper, General Electric Company, Coshocton, OH, November 29, 2001.

[27] Adams-Melvin, B. L., et al, "Effects of Copper Foil Type and Surface Preparation on Fine Line Image Transfer in Primary Imaging of Printed Wiring Boards," *P.C. World Convention VII*, Basel, Switzerland, May 21–24, 1996.

[28] GE Electromaterials, "GTEK® Product Data: HTE Copper Foil," Applications Note, General Electric Company, Coshocton, OH, November 29, 2001.

[29] Institute for Interconnecting and Packaging Electronic Circuits, "Generic Standard on Printed Board Design, IPC-CF-150E," Northbrook, IL, February 1998, http://www.ipc.org.

[30] Dietz, K., "Fine Lines in High Yield (Part LXXXII): Fighting the Etch Factor and Etch Non-Uniformity," *CircuitTree*, July 1, 2002.

[31] Monroe, S., and O. Buhler, "The Effects of Etch Factor in Printed Wiring Characteristic Impedance," *IEEE 11th Annual Regional Symposium on EMC*, Northglenn, CO, October 3, 2001.

[32] Staniforth, A., and M. Gaudion, "The Effects of Etch Taper, Prepreg and Resin Flow on the value of the Differential Impedance," Application Note AP148, Polar Instruments, Ltd., 2002.

[33] Brist, Gary, et al. "Reduction of High-Frequency Signal Loss Through the Control of Conductor Geometry and Surface Metallization," *SMTA International*, September 22, 2002.

[34] Rowland, R., "Substrate Trends and Issues," *SMTA International 2002 Conference*, Rosemont, IL, September 24–26, 2002.

[35] Parquet D., and D. Boggs, "Alternatives to HASL: Users Guide for Surface Finishes," Applications Note, Merix Corp., Forest Grove, OR.

[36] Merix Corp., "Soldermasks," Technical Brief, Forest Grove, OR, June 1997.

[37] Mitchell, C. E., "Photoimagable Solder Mask: the Case for UV Blocking Laminate," Technical Paper, GE Electromaterials, General Electric Company, Coshocton, OH, February 1993.

CHAPTER 2
Resistance of Etched Conductors

2.1 Introduction

Conductor resistance is the dominant loss factor in PWB traces below roughly 1 GHz for many common laminates, including FR4. The trace width, thickness, ambient temperature, signal frequency content, and the proximity to other conductors and to its return all proportionally play a roll in determining the conductors' loop resistance (and thus the conductor's loss). The increase in resistance at high frequencies due to skin effect is especially important because it's one of the factors that cause unequal attenuation of each harmonic making up a signal. As discussed in Chapters 5 and 7, frequency-dependent attenuation contributes to signal dispersion and distortion.

Additionally, determining the dc resistance of a trace is important when working with termination schemes that draw dc (such as Thevinin terminations) because if not managed properly, the dc voltage drop can alter bias levels.

This chapter first addresses resistance at low frequencies (Section 2.2) and discusses loop resistance and the proximity effect (Section 2.3). In so doing, it introduces the resistance matrix (a prelude to the capacitance and inductance matrices presented in Chapters 3 and 4). Simple hand calculations to compute the skin depth and then the increase in resistance with frequency due in both the conductor and its return are presented in Sections 2.4 and 2.5. Section 2.6 discusses how the surface roughness of a trace increases resistance at high frequencies.

2.2 Resistance at Low Frequencies

Figure 2.1 shows a copper bar on an insulator placed over a copper sheet. The dc resistance from one end of the bar to the other is found by (2.1):

$$R_{dc} = \frac{\rho}{A} length \qquad (2.1)$$

where ρ is the material's volume resistivity—a proportionality factor (with units of ohm-meter) that determines the current per unit area (the current density) that flows when a given electric field is applied. Table 2.1 lists resistivity at room temperature for several metals. The conductor cross sectional area (A) determines the amount of metal supporting the current flow along the conductor's length.

At low frequencies, the entire cross sectional area of the conductor is available to carry current. For a rectangular conductor such as most PWB traces, area (A) is

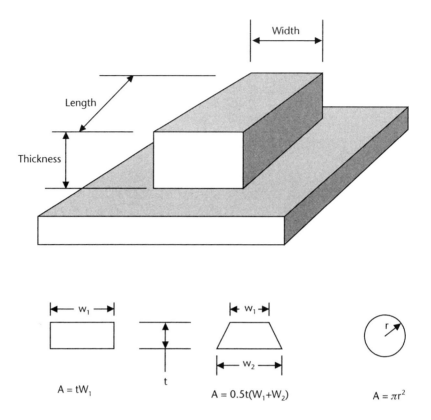

Figure 2.1   Dimensions of a metal bar on an insulator over conductive sheet.

Table 2.1   Resistivity and Temperature Coefficient for Metals

Metal	ρ (ohm-meter, 25°C)	ρ (ohm-inch, 25°C)	α
Silver	$1.59 - 1.62 \times 10^{-8}$	$403.8 - 411.4 \times 10^{-12}$	0.0038
Copper	1.76×10^{-8}	447.0×10^{-12}	0.0043
Gold	2.40×10^{-8}	60.96×10^{-12}	0.0034
Aluminum	2.83×10^{-8}	71.88×10^{-12}	0.0039–0.0043
Brass	$7-8 \times 10^{-8}$	$177.8–203.2 \times 10^{-12}$	0.001–0.002
Tin	11.5×10^{-8}	292.1×10^{-8}	0.0042

therefore the product of the conductors' thickness (t) and its width (w) at dc. The terms for circular wire and trapezoidal trace (as is sometimes obtained on a PWB due to over etching) are also shown.

Resistance of rectangular half-ounce ($t = 0.65$ mils) and one-ounce ($t = 1.4$ mils) copper trace are plotted for various widths in Figure 2.2, at 25°C. A rough general rule is that the room temperature dc resistance is about 700 mΩ/inch length for

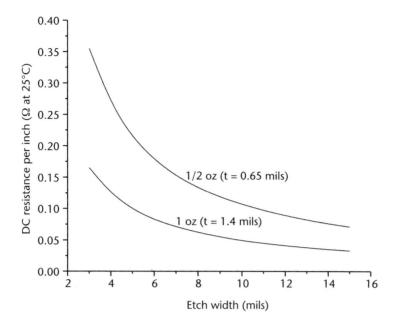

Figure 2.2   Room temperature dc resistance of rectangular half- and one-ounce copper trace.

a 1-mil-wide half-ounce copper and half that for one-ounce copper. Using that approximation and recalling that resistance decreases as width increases, a 5-mil-wide half-ounce trace has a dc resistance of 140 mΩ per inch [versus 138 mΩ from (2.1)].

Resistivity is specified at a reference temperature, but as shown in (2.2) the metal's resistance increases linearly with increasing temperature. The rate of change is determined by the temperature coefficient (α). The difference between the reference temperature used to specify ρ (usually 25°C) and the temperature of interest is signified by Δt. The resistance at the reference temperature is multiplied by the factor R_m given in (2.2) to determine the resistance at any other temperature:

$$R_m = 1 + (\alpha \Delta t) \tag{2.2}$$

Equation (2.2) is valid over the temperature range experienced by commercial and industrial PWBs and is plotted in Figure 2.3. Higher-order terms are required when working with very high or low temperatures.

Table 2.1 [1, 2] lists the resistivity at 25°C and temperature coefficients for some common metals. The metals' purity and processing greatly affect both the resistivity and the temperature coefficient, so the values in Table 2.1 should be regarded as approximate.

Example 2.1

    (a) What is the room temperature dc resistance of a 10-in long, 1-oz, 5-mil-wide copper microstrip, as shown in Figure 2.1?

    (b) What is it at an ambient of 60°C?

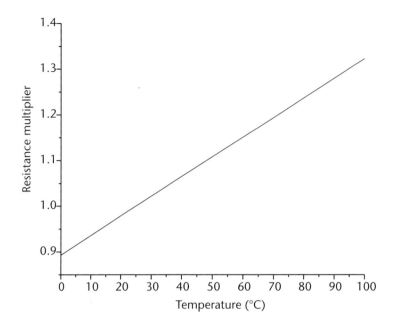

Figure 2.3 Temperature correction factor for copper.

Solution

(a) Assuming the conductor is rectangular, and that a 1-oz copper trace is 1.4 mils thick, (2.1) gives the resistance at 25°C (room temperature) as: $R = \dfrac{\rho \times length}{A} = \dfrac{1.76 \times 10^{-8} \times 0.254}{35.56 \text{ um} \times 127 \text{ um}} = 0.99\Omega$.

(b) For a 60°C ambient, $\Delta t = 60 - 25 = 35$°C. Therefore, from (2.2) and Table 2.1, $R_m = 1 + (\alpha \Delta t) = 1 + (0.0043 \times 35) = 1.15$. That is, at 60°C the trace resistance will increase 15% above the room temperature value to $(1.15 \times 0.99\Omega) = 1.14\Omega$.

2.3 Loop Resistance and the Proximity Effect

The resistance between the ends of a copper trace was computed in Example 2.1, but the resistance in the return path must also be included to properly compute total conductor loss. For the general case of a single wire and its return, the total loop resistance is given in (2.3), where R_{11} is the total loop resistance (its *self resistance*), R_e is the end-to-end resistance of just the trace, and R_r is the resistance of the return path.

$$R_{11} = R_e + R_r \tag{2.3}$$

Using the results from Example 2.1, R_e is ~1Ω, and assuming the return is 0.25Ω, R_{11} is therefore 1.25Ω under dc conditions. Forcing 1A down this trace causes a 1-V drop across the trace, plus an additional 0.25-V drop across the return.

2.3.1   Resistance Matrix

Calculating the loop resistance becomes more involved when multiple signals use the same return path to complete their loops. For example, three identical 10-in-long, 5-mil-wide, one-ounce microstrips are shown in Figure 2.4, all shorted at their ends to a common point on the return plane by zero-ohm jumpers. From the previous discussion, R_{11} is 1.25Ω at room temperature.

If 1-A dc is forced down a single trace, the total voltage drop will be R11 × 1A = 1.25V. However, if the other two traces also conduct 1A in the same direction, the voltage drop across each trace remains 1V while the voltage drop across the return path increases threefold to 0.75V. The total loop voltage drop simultaneously experienced by any of the traces is therefore 1.75V. From the perspective of a single line, it appears as if its resistance has increased 40% strictly as a result of the current drawn by its neighbors. Because of the common return path, the lines are said to share *mutual resistances,* and a resistance matrix may be formed as in (2.4) to show the interaction between conductors and the return paths.

$$R = \begin{matrix} R_{11} & R_{12} & R_{13} \\ R_{21} & R_{22} & R_{23} \\ R_{31} & R_{32} & R_{33} \end{matrix} \qquad (2.4)$$

The values appearing on the main diagonal (R_{11}, R_{22}, R_{33}) represent the loop self resistance of conductors 1, 2, and 3, including its return path resistance, per (2.3) when all other conductors have zero current.

The terms off the main diagonal represent the mutual resistance appearing between each conductor. For example, R_{12} is the mutual resistance between conductor 1 and 2. Naturally this is identical to the mutual resistance when measured from

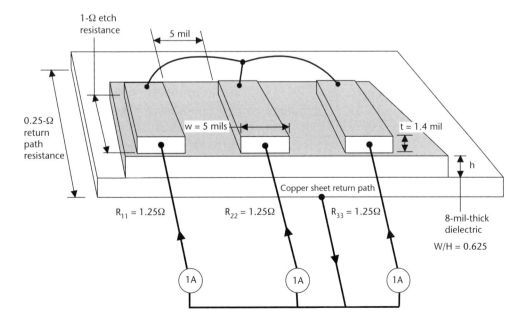

Figure 2.4   Three microstrips sharing a common return.

conductor 2 to 1 (R_{21}), and the terms off the main diagonal can be omitted without loss of information.

The IR drop, including mutual resistance effects, is found by multiplying the resistance matrix by a matrix representing the current flow, yielding a voltage matrix (2.5):

$$V = RI \qquad (2.5)$$

The resistance matrix for the three-conductor example of Figure 2.4 is shown in (2.6):

$$R = \begin{matrix} 1.25 & 0.25 & 0.25 \\ 0.25 & 1.25 & 0.25 \\ 0.25 & 0.25 & 1.25 \end{matrix} \qquad (2.6)$$

Multiplying (2.6) by a current matrix having 1A in each conductor yields the voltage matrix (2.7):

$$V = \begin{matrix} 1.25 & 0.25 & 0.25 \\ 0.25 & 1.25 & 0.25 \\ 0.25 & 0.25 & 1.25 \end{matrix} \times \begin{matrix} 1 \\ 1 \\ 1 \end{matrix} = \begin{matrix} 1.75\,V \\ 1.75\,V \\ 1.75\,V \end{matrix} \qquad (2.7)$$

As expected, this is the same result as that obtained previously. Of course, in practical systems the switching current will be in the milliamp range, but using 1A as an excitation current in (2.7) makes it convenient to scale the results to any current. For example, the voltage drop is 1.75 mV if the conductors switch 1 mA each.

2.3.2 Proximity Effect

At low frequencies the current is uniformly distributed throughout the conductor, but at high frequencies the current tends to migrate to the surface of the conductor that is facing the return path. This concentration of current on adjacent surfaces is often called the *proximity effect* [3, 4], and in stripline or microstrip is responsible at high frequencies for the (gradual) concentration of current under the trace in the ground or power plane that acts as the signals return. As described in Section 2.5, this results in the return current spreading out and using the entire return path at low frequencies (the path of lowest resistance), while at high frequencies the effect is for the return path current to collect underneath the trace. This increases the loop resistance (thereby increasing conductor loss) but minimizes its inductance (path of smallest area).

For example, the resistance matrix for the 10-in-long three-conductor system of Figure 2.4, measured at 1 GHz, is shown in (2.8).

$$R = \begin{matrix} 12.82 & 1.04 & 0.285 \\ 1.04 & 12.82 & 1.04 \\ 0.285 & 1.04 & 12.82 \end{matrix} \qquad (2.8)$$

The concentration of high-frequency return current underneath a signal trace causes the off diagonal terms in (2.8) to be unequal. In fact, the mutual resistances fall as a function of distance because the return currents of distant conductors are less able to commingle with the return currents of immediate neighbors. In comparison, the mutual terms in (2.7) are all equal because at dc the current spreads out evenly across the entire width of the return plane, and each conductor is equally able to interferer with all others. This is illustrated in Figure 2.5, which shows an edge view of the three conductors and the current density in the return plane.

In (2.7) and (2.8), the self resistances are identical in their respective matrices because the traces are all the same size and height above the return path. Skin effect and proximity effect makes the self resistance in (2.8) higher than that appearing in (2.7).

As in the dc case, the voltage drop experienced by each conductor can be found at high frequency by multiplying the resistance matrix by a current matrix representing the switching condition of interest. The results are shown in (2.9) for a 1-A current in each conductor:

$$V = RI = \begin{array}{ccc} 12.82 & 1.04 & 0.285 \\ 1.04 & 12.82 & 1.04 \\ 0.285 & 1.04 & 12.82 \end{array} \times \begin{array}{c} 1 \\ 1 \\ 1 \end{array} = \begin{array}{c} 14.15 \\ 14.90 \\ 14.15 \end{array} \qquad (2.9)$$

Because the resistance matrix was obtained at 1 GHz, (2.9) shows the voltage drop per amp at that frequency. The frequency-dependent nature of the resistance makes the results in (2.9) significantly higher than that at dc (2.7).

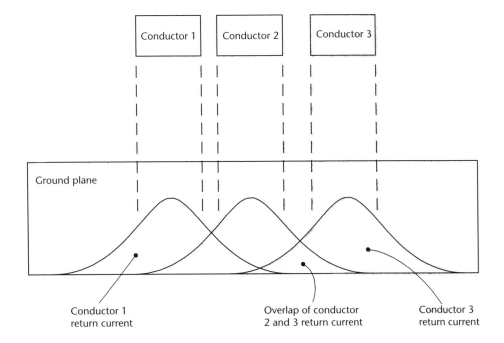

Figure 2.5   Return plane current density for three closely spaced conductors.

As expected, conductors 1 and 3 have the lowest (but identical) voltage drop (14.15V), while conductor 2 has the highest drop. This shows that when all three lines simultaneously drive current in the same direction, the loop resistance of the center conductor is higher than that of the outer loops. As loss is a function of conductor resistance, it's clear from (2.9) that the switching activity of neighboring traces can adversely affect loss of adjacent signals. This leads to frequency- and data-dependent losses that can cause pulse distortion and may appear as dispersion and intersymbol interference (ISI), topics covered in Chapter 7. Ways to mitigate this are discussed in Chapters 8 and 9.

The return path need not be confined to the power or ground planes forming the microstrip or stripline structure. Adjacent traces can act as returns, and the effects of signals switching in the opposite direction (or not switching at all) can be determined by appropriately setting the current matrix. For example, multiplying (2.8) with a current matrix of 1A in one conductor and zero amps in the remaining two yields a loop voltage drop of 12.8V. However, setting the two loops to −1A (signifying switching in the direction opposite of the first conductor) yields a loop voltage of 11.5V. This is lower because part of the return current has been removed from the return plane by the oppositely switching signals.

2.4 Resistance Increase with Frequency: Skin Effect

The migration of current from the inner portions of the conductor to the surface occurs gradually as frequency increases. It is called the *skin effect* because it's as if the current is traveling in a thin layer (skin) near the conductor's surface. The thickness of this layer is called the *skin depth* or the *depth of penetration* and for nonferrous metals is given by (2.10).

$$\delta = \sqrt{\frac{\rho}{f\pi\mu_0}} \tag{2.10}$$

where ρ is the materials resistivity (Table 2.1) and μ_0 is the permeability in a vacuum (exactly equal to $4\pi 10^{-7}$ F/m, or about 31.92 nH/in).

For copper at room temperature, with the frequency in megahertz, (2.10) becomes:

$$\delta = \sqrt{\frac{4.458 \times 10^{-9}}{f(\text{MHz})}} \tag{2.11}$$

The current actually penetrates exponentially into the conductor and does not abruptly stop at a boundary equal to one skin depth. In fact, at one skin depth the field strength is $\dfrac{1}{e^{n=1}} = 36.8\%$ of what it is at the surface [5], and it requires 5 times ($n = 5$) a skin depth to fall to under 1% of the surface value.

Example 2.2

What is the penetration depth of a copper trace at 10 MHz and 100 MHz?

Solution

Using (2.11) at 10 MHz, $\delta = \sqrt{\dfrac{4.458 \times 10^{-9}}{10}} = 21.14 \times 10^{-6}\, m (= 0.83\ \text{mils})$; at 100 MHz $\alpha = 6.68 \times 10^{-6}\,m$ (0.26 mils).

Figure 2.6 shows the skin depth in copper across frequency at room temperature and at 75°C. Half- and one-ounce copper thicknesses are illustrated for reference.

Half-ounce copper is nominally 0.65 mils thick: Figure 2.6 shows that at 10 MHz, the trace is fully penetrated, so the entire cross-sectional area of the conductor is available to carry the current. Therefore, the resistance of the trace itself is essentially the same as the value at dc.

However, at 100 MHz, the signal acts as if it is confined to a sheet only 0.26 mils below the surface. Because the entire conductor area is not used at high frequencies, one would expect the resistance to be higher then that measured at low frequencies (such as dc), and this is indeed the case. In fact, as shown in (2.10), the resistance increases as $\sqrt{f}$ for frequencies above the skin effect onset frequency.

The lines in Figure 2.7 show the loop resistance as defined in (2.3) of a single rectangular microstrip (plotted with a dashed line) and stripline (solid line) made from a half- and one-ounce 5-mil-wide copper trace across frequency. The $\dfrac{w}{h}$ ratio equals 1.7 for both the microstrip and stripline traces, and the stripline is equally spaced between the two return planes. As is discussed in Chapter 9, h is one of the factors in determining the impedance of a trace. In fact, assuming FR4, the microstrip traces depicted in Figure 2.7 will have an impedance of just over 50Ω, while the striplines impedance is in the lower 30-Ω region.

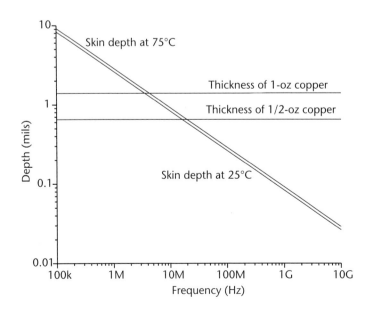

Figure 2.6   Skin depth for copper trace at 25°C and 75°C.

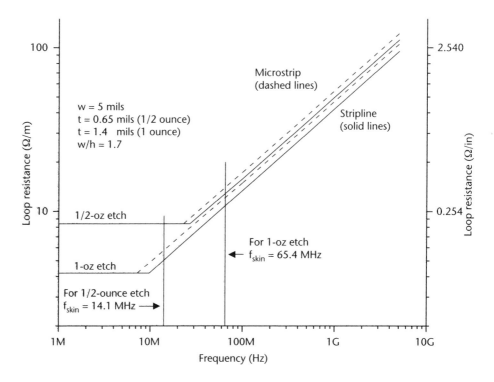

Figure 2.7   Stripline and microstrip loop resistance.

The frequency and resistance are plotted on logarithmic scales. As shown, the half-ounce trace retains is dc value to about 20 MHz, where it then transitions to the $\sqrt{f}$ behavior. Being thicker, the one-ounce trace transitions sooner (roughly at 7 MHz) because it can only be fully penetrated when the skin depth is large, and that occurs at a lower frequency.

For nonferrous metal bars that are wider than they are thick, the transition from the dc to the $\sqrt{f}$ behavior is conservatively estimated by (2.12) [6]:

$$f_{skin} = \frac{4\rho}{\pi \mu_0 t^2} \tag{2.12}$$

Equation (2.12) computes the frequency where the skin depth is half the trace thickness.

The value for f_{skin} as computed by (2.12) is shown in Figure 2.7 for both thicknesses of trace. In both cases, the intersection of the horizontal line representing the dc resistance and the sloping line representing the ac resistance occurs at a lower frequency than predicted by (2.12), showing that the $\sqrt{f}$ behavior is well established by the frequency predicted by (2.12). Figure 2.7 implies an abrupt transition between these two regions, but in fact the transition is gradual.

For a rectangular copper trace at room temperature, (2.12) reduces to (2.13):

$$f_{skin} = \frac{s}{t^2} \tag{2.13}$$

where f_{skin} is approximately the frequency (in Hertz) where the resistance begins to increase as the $\sqrt{f}$, and $s = 17.83 \times 10^{-3}$ for thickness (t) in meters. Alternatively, $s = 27.64$ if the thickness is expressed in mils. In that case (2.13) reports f_{skin} in MHz.

Example 2.3

At approximately what frequency do 1-oz and half-ounce copper traces show an $\sqrt{f}$ increase in resistance?

Solution

From (2.13), for half-ounce copper $F_{skin} = \dfrac{s}{t^2} = \dfrac{27.64}{0.65^2} = 65.4$ MHz; it's 14.1 MHz for one ounce. That is, on one-ounce copper, frequencies greater than about 14 MHz will experience trace resistance increasing as $\sqrt{f}$. This means that a signal's harmonics (constituent frequency components) that are higher than about 14 MHz will be attenuated unequally, with the attenuation increasing with frequency. As described in Chapters 5 and 7, this leads to signal distortion.

2.5   Hand Calculations of Frequency-Dependent Resistance

It's possible to properly hand calculate skin effect resistance for round wires, especially if no other wires are nearby [7]. However, because at high frequencies the current distribution is not uniform in rectangular conductors, it's much more difficult to hand calculate high-frequency resistance for PWB trace. This is especially so in the presence of other conductors. At very high frequencies, the current tends to concentrate on the surface nearest the return path due to the proximity effect and peaks in the corners of the trace [7], but the hand calculations assume the electric and magnetic fields are uniform along the conductor's width [7, 8]. This failure to properly account for the irregular current concentration can cause these calculations to underestimate the trace resistance by up to 50% [9].

Additionally, at high frequency the current distribution in the return path is also difficult to estimate by hand, especially when multiple lines share a common return path. For wide traces close to the return path, the return current concentrates directly under the trace. However, for narrow traces high above a return path, it tends to spread out beyond the width of the trace [8]. At high frequency, the ratio of a microstrip's width to height above the return path (w/h) determines the amount of spreading: only when w/h exceeds three to five is the return current more or less confined underneath the trace [10, 11]. Thus, the return currents will tend to concentrate directly underneath a 5-mil-wide 30-Ω microstrip on FR4 ($\varepsilon_r = 4$; $w/h \sim 5$) and will show some spreading for a 50-Ω trace ($w/h \sim 2.5$), but will spread out well beyond a 5-mil-wide 75-Ω trace ($w/h \sim 1$) (see Figure 2.8).

Field-solving software should be used to properly determine the magnitude of frequency-dependent resistance. This is especially true when multiple lines having a small w/h ratio share a common return, as the return currents for these type lines will tend to overlap, effectively increasing the loop resistance. This is visible in the resistance matrix (2.8) appearing in Section 2.3.2 for a 5-mil-wide line 8 mils above a return ($w/h = 0.625$).

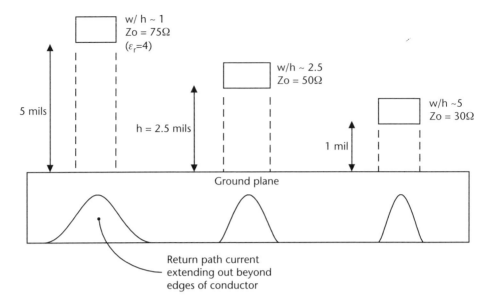

Figure 2.8   Current spreading in return path for microstrips of various heights.

Nonetheless, provided the limitations are understood, simple hand calculations that approximate the resistance are valuable during initial engineering studies or when using a field solver for the first time.

2.5.1   Return Path Resistance

An approximation formula to compute the ac resistance of the ground plane underneath a single microstrip is given in [12]. It can be recast as (2.14):

$$Rgnd = 0.55 Rdc \left(\frac{t}{\delta}\right)\left(1 - e^{-\frac{w'}{1.2\pi}}\right) \tag{2.14}$$

where $w' = \dfrac{w}{h}$ (the trace width over its height above the return path), δ is the skin depth given in (2.10) or (2.11), and t is the trace thickness.

2.5.2   Conductor Resistance

Neglecting return path resistance, the resistance for an isolated stripline or microstrip at frequencies above f_{skin} can be approximated by (2.15) [6]:

$$R_{ac} = R_{dc}\sqrt{\frac{f}{f_{skin}}} \tag{2.15}$$

This calculation provides a way to easily estimate the resistance of the trace only and becomes increasingly accurate as frequency exceeds f_{skin}. The resistance contribution of the return path must be included to determine the overall loop resistance (and so, to properly estimate the total losses).

If the resistance is known at a frequency f_1 (2.16) shows how to determine the resistance at some other frequency (f_2), provided both are above f_{skin}:

$$R_{ac2} = \sqrt{\frac{f_2}{f_1}} \qquad (2.16)$$

For example, the resistance at 1 GHz would be 3.16 times greater than the resistance at 100 MHZ.

2.5.3   Total Loop Resistance

The total loop resistance can be estimated for microstrip (and, stripline to within about 25%) by combining the results of (2.15) and (2.14):

$$R_{loop} = R_{ac} + R_{gnd} \qquad (2.17)$$

The results from (2.17) have increasing error with respect to field-solving software for larger w/h. For lower impedances using one-ounce trace, (2.17) underreports the loop resistance of narrow microstrip by between less than 5% ($w/h = 1$) and 10% ($w/h = 5$). Calculations with half-ounce trace show at least twice the error, and error increases as width increases, regardless of copper thickness.

Although (2.14) and (2.15) were created to compute microstrip resistances, (2.17) can be used to predict the loop resistance of *symmetrical stripline* (where the trace is equidistant from the bottom and top plates). Because the return paths are assumed to be equally effective in returning current, R_{gnd} calculated by (2.14) is divided in half. Nonetheless, the loop resistance will be underpredicted by as much as 50% for small w/h.

2.6   Resistance Increase Due to Surface Roughness

The grain and surface roughness of the copper trace become increasingly significant in determining conductor resistance as the current migrates to the conductor's surface at high frequencies. The peaks and valleys of the rough surface extend the mean free path the electrons must travel and thus increase the resistance over that predicted by a simple $\sqrt{f}$ relationship.

Unfortunately, data that describes increasing resistance explicitly due to surface roughness of copper foils is not abundant in the literature, but data discussing conductor loss is prevalent. As discussed in Chapter 7, work has been done to develop empirical formulas to account for surface roughness in microstrip conductor loss calculations without explicitly computing the loop resistance.

However, measurement data is presented in [13] for the increase in resistance at 3 GHz due to surface roughness as a function of skin depth for several metals, including copper. This data is in reasonable agreement with that found in [14] for similar values of average surface roughness (Ra). It's possible to calculate the expected approximate resistance increase for electrodeposited and rolled copper foils at 3 GHz from the data appearing in [13, 14] and in Chapter 1. Doing so suggests that half- and one-ounce rolled copper each show about a 30% increase in high-frequency resistance due to surface roughness, while half-ounce electrodeposited copper shows about a 40% increase. One-ounce electrodeposited shows about a 50% increase over that predicted by the $\sqrt{f}$ relationship.

2.7 Summary

A trace's resistance, including that of its return path, is an important factor in determining signal loss. At high frequencies, the skin effect causes the resistance to increase as $\sqrt{f}$ over the dc value predicted by (2.1). The onset of skin effect is conservatively predicted by (2.12).

A metal's temperature coefficient will cause resistance to increase with temperature [from (2.2)], and return current commingling in a return path (such as a ground plane) will cause an apparent increase in trace resistance due to the proximity effect (described in Section 2.3). All of these things may be compactly summarized in a resistance matrix [such as that presented in (2.4)].

References

[1]   Hammond, P., *Electromagnetism for Engineers*, 3rd Ed., Oxford: Pergamon Press, 1986.

[2]   The Chemical Rubber Company, *CRC Handbook of Chemistry and Physics (1986–1987)*, 67th Ed., Cleveland, OH: CRC Press, 1987, pp. E–9.

[3]   Johnson, W. C., *Transmission Lines and Networks*, Chapter 3, New York: McGraw Hill, 1950.

[4]   Grivet, P., *The Physics of Transmission Lines at High and Ultra High Frequencies, Vol. 1*, London: Academic Press, 1970.

[5]   Ramo, S., et al., *Fields and Waves in Communication Electronics*, 3rd Ed., New York: John Wiley and Sons, 1994, p. 151.

[6]   Vu Dinh, T., et al., "New Skin-Effect Equivalent Circuit," *Electronics Letters*, Vol. 26, No. 19, September 13, 1990, pp. 1582–1584.

[7]   Paul, C., *Analysis of Multiconductor Transmission Lines*, New York: John Wiley and Sons, 1994.

[8]   Bertin, C. L., "Transmission-Line Response Using Frequency Techniques," *IBM Journal of Research and Development*, January 1964, pp. 52–63.

[9]   Paul, C., *Analysis of Multiconductor Transmission Lines*, New York: John Wiley and Sons, 1994, p. 177.

[10] Faraji-Dana, R., and Y. L. Chow, "The Current Distribution and AC Resistance of a Microstrip Structure," *IEEE Journal of Microwave Theory and Techniques*, Vol. 38, No. 9, September 1990, pp. 1268–1277.

[11] Pucel, R. A., et-al., "Losses in Microstrip," *IEEE Trans. Microwave Theory and Techniques*, Vol. MTT-16, No. 6, June 1968, pp. 342 – 350; also see corrections in MTT-16, No. 12, Dec 1968, pp. 1064.

[12] Faraji-Dana, R., and Y. L. Chow, "The Current Distribution and AC Resistance of a Microstrip Structure," *IEEE Journal of Microwave Theory and Techniques*, Vol. 38, No. 9, September 1990, p. 1273.

[13] Saad, T. S., *Microwave Engineers Handbook, Vol. 2*, Norwood, MA: Artech House, 1971, p. 186.

[14] Tanka, H., and F. Okada, "Precise Measurements of Dissipation Factor in Microwave Printed Circuit Boards," *IEEE Trans. Inst and Meas.*, Vol. 38, No. 2, April 1989, pp. 509–514.

CHAPTER 3

Capacitance of Etched Conductors

3.1 Introduction

This chapter discusses the way in which capacitance is formed on PWBs. Capacitance is one of the fundamental circuit elements forming a transmission line, and capacitive coupling between PWB etch contributes to noise voltages.

The relationship between capacitance and charge is reviewed in Section 3.2, which naturally leads to the introduction of the dielectric constant. This is used in the parallel plate capacitor discussion appearing in Section 3.3, which focuses on the limitation of the parallel plate capacitor model for computing etch capacitance. As discussed, these formulas are not suited for computing microstrip and stripline capacitance. Capacitance and impedance formulas specifically suited for etched conductors are presented in Chapter 9.

Mutual capacitance contributes to crosstalk, jitter, and ISI, and is discussed in Section 3.5. Section 3.5 shows how to interpret the capacitance matrix presented by many field solvers, and it shows how to obtain the mutual and self capacitance of an etch. This background is a fundamental introduction for the crosstalk and differential impedance material presented in Chapter 8.

Dielectric losses contribute to signal degradation at very high signaling rates. The loss tangent is introduced in Section 3.6 and represents lossy dielectrics as a shunt conductance. This is an important prelude to the lossy transmission line model discussed in Chapter 5.

The dielectric characteristics of some laminates are presented in tabular form later in the chapter, and Section 3.7 discusses the effects temperature and humidity has on FR4 type epoxy resins. This material is useful when developing transmission line circuit models.

In most cases, detailed proofs are not offered for the mathematical relationships presented here. This is in keeping with this book's spirit, which presents the mathematical results and uses physical intuition to explain concepts. Numerous citations are present in the reference for those wishing to delve deeper into the mathematics.

3.2 Capacitance and Charge

Charge accumulates between two conductors separated by a dielectric when a voltage differential is present between them. The magnitude is directly proportional to the differential voltage. Capacitance (C, units of farads) is the proportionality factor relating the accumulation of charge (Q, in Coulombs) to the voltage difference applied between the plates (V, in volts), as shown in (3.1):

$$Q = CV \qquad (3.1)$$

Holding capacitance constant, differentiating (3.1) with respect to time and recognizing that current is the rate at which charge changes with time yields the familiar equation relating voltage, current, and time to capacitance in (3.2):

$$Cdv = idt \qquad (3.2)$$

The utility of (3.2) in determining the current necessary to charge a capacitive load is illustrated in Example 3.1.

Example 3.1

What constant current is required to charge a 10-pF capacitor from 0V to 1.5V in 1 ns?

Solution

The capacitor will have a change in voltage of (1.5V − 0V) = 1.5V, and it will obtain this difference in voltage in 1 ns. This makes dv = 1.5V and dt = 1ns.

Therefore, from (3.2), the average current is $i = \dfrac{Cdv}{dt} = \dfrac{10 \text{ pF} \times 1.5 \text{V}}{1 \text{ ns}} = 15$ mA.

Notice that this same current value is required to charge the capacitor between any two voltage values separated by 1.5V (for example, charging from 3V to 4.5V).

3.2.1 Dielectric Constant

The capacitance of a parallel plate capacitor is proportional to the plate area and inversely proportional to the plate's separation. The *permittivity* is a proportionality factor described by Coulomb's law, which can be used to relate capacitance to the plate area and spacing (for example, see [1]). In free space, it has a value of ε_0 = 8.854×10^{-12} F/M (or 224.9×10^{-15} F/in).

Usually the permittivity of a dielectric is given with respect to (i.e., relative to) the permittivity of free space, as show in (3.3):

$$\varepsilon_r = \frac{\varepsilon_{material}}{\varepsilon_0} \qquad (3.3)$$

where ε_r is the *relative permittivity*. However, circuit board and capacitor manufacturers favor the term *dielectric constant* (abbreviated *Dk* or sometimes just *K*). The more formal term ε_r is used in this text, even though it's more cumbersome. The two terms are identical.

Table 3.1 and Table 3.2 (which appears in Section 3.6) list the relative permittivity of some common materials and PWB laminates at 25°C. The tables show ε_r followed by the loss tangent (discussed in Section 3.6.2). For example, at 1 MHz distilled water has a dielectric constant ε_r of 78.2 and a loss tangent of 0.04. As discussed in Section 3.7, water uptake by laminates changes their dielectric constant and loss values.

Some of the materials (such as polystyrene) show essentially no change in ε_r across frequency, while the other materials do show a reduction as frequency increases. In Table 3.1 the values for barium titanate (used as a dielectric in many ceramic capacitors) show the largest change and (unlike the other listed materials) will also show a change in ε_r with the strength of the applied electric field. As discussed in Chapter 10, this results in the voltage variability of capacitance common in most ceramic capacitors.

3.3 Parallel Plate Capacitor

The electric field lines appearing between the plates of a parallel plate capacitor are illustrated in Figure 3.1.

If the plates are close enough together (or if the plates are very wide), the charge on the plates perimeter contributes only a small portion of the capacitors total charge. However, as the plate width decreases or the plate separation increases *fringing fields* increasingly contribute to the total capacitance, and their contribution must be included to properly compute capacitance.

Neglecting fringing, the capacitance for a parallel plate capacitor is given in (3.4):

$$C = \frac{\varepsilon_r \varepsilon_0 lw}{h} \tag{3.4}$$

where l and w represent the area of one capacitor plate, and h represents the separation between the two plates.

Because it neglects fringing, (3.4) will underreport the capacitance of microstrip or stripline unless w/h is extremely large. In fact, the w/h ratio must exceed roughly

Table 3.1   Relative Permittivity/Loss Tangent for Common Materials

Material	1 MHz	100 MHz	3 GHz (1 GHz)	Notes
Balsawood	1.37/0.012	1.30/0.0135	1.22/0.10	
Barium titanate	1,100/0.0002		600/0.0023	
Distilled water	78.2/0.04	78/0.05	76.7/0.157	
E-glass	6.4/0.0018		(6.13/0.0039)	Measured at 1GHz rather than 3 GHz
Polycarbonate (Lexan®)	3.0/0.010			Registered Trademark of GE Plastics
Polyester (Mylar®)	3.0/0.016		2.8/0.018	Registered Trademark of E.I. duPont de Nemours & Co., Inc.
Polystyrene	2.55/0.0002	2.55/0.0004	(2.55/0.0004)	Measured at 1 GHz, rather than 3 GHz
PTFE (Teflon)	2.1/0.0002	2.1/0.0002	2.1/0.0002	Registered Trademark of E.I. duPont de Nemours & Co., Inc.

Sources: [2–7].

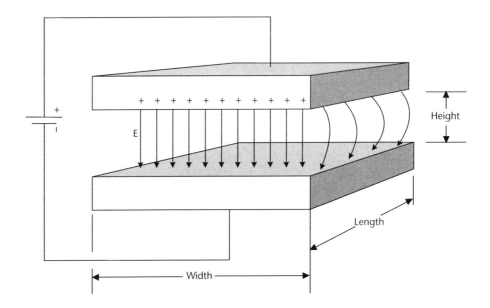

Figure 3.1 Electric fields between plates of a parallel plate capacitor.

50 before the error falls under 10% [8], making (3.4) grossly in error in predicting etch capacitance unless the etch impedance is extremely low (well under 20Ω). However, (3.4) is useful when computing capacitance between wide power/ground planes in circuit boards, as such structures have a large w/h ratio and so fringing is less of a factor.

An equation for computing the capacitance of rectangular parallel plate capacitors that does include a fringing correction factor is given in [9], while [10] presents equations for a capacitor with circular plates, and [11] shows how to compute the capacitance of a cylinder. As our focus is on etched conductors, these results are not presented here. Chapter 9 presents equations for microstrip and stripline capacitance and impedance. Alternatively, the reciprocity principal (presented in Section 4.2.5) may be used to compute capacitance if inductance is known.

Example 3.2

A circuit board 9 inches long by 6 inches wide is fabricated on FR4 (ε_r = 4.5 at 100 MHz). Assume vias have a 30-mil antipad size and on average the board has a density of 150 vias per square inch. At 100 MHz, what is the capacitance appearing between a power and ground plane separated by 5 mils (a) neglecting vias and (b) including vias?

Solution

(a) The planes form the two plates of a parallel plate capacitor. Equation (3.4) is applicable because the width-to-height ratio is very large, so fringing in the air will not significantly contribute to the capacitance. Accordingly, using engineering units, the capacitance appearing between the planes is:

$$C = \frac{\varepsilon_r \varepsilon_0 lw}{h} = \frac{4.5 \times 224.9 \text{ fF} \times 9 \times 6}{0.005} = 10.9 \text{ nF} \tag{3.5}$$

(b) The antipads act as metal punch outs from the power/ground planes, reducing the total plate area forming the capacitor. With a 30-mil diameter, each antipad has an area of 0.707 mils^2. With an average via density of 150 vias per in^2 and a board area of 54 in^2, the vias will remove an average of 5.73 in^2 of metal from the plates, representing a loss of 10.6% in the total plate area. Because (3.4) is linear with plate area, the capacitance from (3.5) will be reduced by 10.6% to 9.8 nF. However, it's worth noting that in practice capacitive coupling from each of the vias to the plane will mitigate this reduction.

A significant decoupling capacitance is seen to occur without using discrete components. As discussed in Chapter 4, the wide power/ground plates results in a low inductance that makes this capacitor especially effective. And, by using laminates with higher ε_r, the capacitance can be increased without any adding additional inductance, further improving the decoupling.

However, as shown in Chapter 8, using laminates with lower ε_r is generally advantageous for signaling. This is because for a given impedance, the lower ε_r results in lower crosstalk and shorter signal propagation time, and for a given etch, impedance allows for a thinner circuit board.

To address this, it's becoming increasingly common for manufacturers to optionally offer circuit boards made with two or more different laminates. This construction creates a board with more than one dielectric constant and is a way to address conflicting signaling and decoupling requirements. On such boards, the layers where adjacent power and ground planes are located have a higher ε_r than the layers used to route signals. In this way, the power plane layers will have higher capacitance than the signal layers, satisfying both the decoupling and signaling requirements. It's also possible to have signal layers consisting of different materials as a way to control cost. In this application, the majority of the PWB consists of lower cost laminate material with a few routing layers of low loss laminate sandwiched in between. A limited number of critical high-frequency signals are routed on one or more of these special low loss layers. This type of construction is more expensive than traditional PWB construction but can be less expensive than a board totally constructed from the low loss laminate. It addresses the inefficiency in using high-performance laminate to route all signals when the laminate is really only required for a few signals.

3.4 Self and Mutual Capacitance

Capacitance is present between any two charged metallic surfaces at different potentials. For a stripline or microstrip PWB etch, this means capacitance can be present from the etch to other etches and to the reference plane. This is illustrated for a three-conductor microstrip in Figure 3.2. The mutual capacitance (C_{12} and C_{23}) expresses the capacitive coupling between conductors. The value of this *coupling capacitance* plays a vital role in determining an etch's switching impedance and the magnitude of crosstalk.

The capacitance from each etch to the reference (ground) plane is defined as C_0. For a given etch, the total capacitance is the sum of the capacitance to the reference plane and to any other conductor having a different potential.

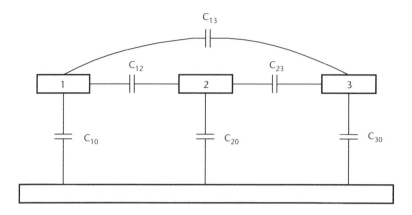

Figure 3.2 Mutual capacitance in a three-conductor system.

As illustrated in Example 3.3, this concept is fundamental to understanding the switching behavior of coupled etches and is discussed in Chapter 8.

Example 3.3

(a) Assume the center conductor in Figure 3.2 changes from 0V to 3.3V while the reference plane and the outer conductors (etches 1 and 3) remain at 0V. What is the total switching capacitance of etch 2?
(b) What is the capacitance of etch 2 if all three conductors simultaneously switch from 0 to 3.3V?

Solution

(a) Once it switches, etch 2 will be 3.3V higher than the ground plane or etches 1 and 3, so charge will have been accumulated on etch 2 relative to the other etches and ground. Figure 3.3 shows the arraignment after etch 2 switches. As shown, the capacitance of etch 2 is the sum of the capacitance to ground (C_{20}) plus the mutual capacitance from etch 2 to etches 1 and 3. This will change the apparent capacitance (and as shown in Chapter 8, the impedance) of etch 2 if the neighboring etches are close enough to make C_{12} and C_{23} large relative to C_{20}. This is an important factor in determining the severity of crosstalk and is numerically examined in Example 3.4.
(b) Because in this case etches 1 and 3 switch at a rate identical to etch 2, there is never a difference in voltage between them and etch 2. The only difference in charge is from conductor 2 to ground. As shown in Figure 3.4, the capacitance is therefore equal to C_{20}.

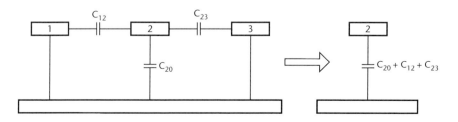

Figure 3.3 Etch 2's capacitance when etches 1 and 3 remain unswitched.

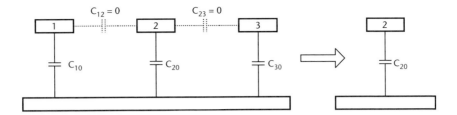

Figure 3.4 Capacitance when three conductors simultaneously switch.

It's implied in Figure 3.2 that the mutual capacitance occurs between the edges of the etches, but this is not so. The mutual capacitance may also include coupling between the tops and bottoms of the etches. As illustrated in Figure 3.5, the quantity of electric field lines reaching the top or bottom surface of another etch depends on the height of the reference planes. The magnitude of the coupling capacitance therefore depends on the etch thickness, its height above the reference plane, and its width (especially in relation to its height above the return plane). Chapter 8 explores this in detail.

3.5 Capacitance Matrix

The relationship between an etch's self capacitance and the mutual capacitance illustrated in Figure 3.2 can be shown with a capacitance matrix. This is somewhat similar to the resistance matrix presented in Chapter 2, but in this case the *self capacitance* given in the capacitance matrix expresses the displacement current that flows from a given conductor to ground when all other conductors are grounded [12]. The *mutual capacitance* terms appear on the off diagonal and represent the charge appearing between a given set of conductors. In circuit terms, this charge is thought of as a coupling capacitance and in the matrix will appear as negative. This is so because for there to be a positive voltage difference between two conductors i and j, conductor j must have a negative charge with respect to conductor i. It follows from (3.2) that a negative charge with a positive voltage difference requires the capacitance to be negative.

The capacitance matrix is represented in (3.6):

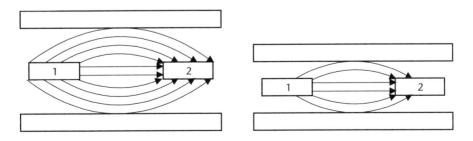

Figure 3.5 Distance to return plane determines the number of field lines reaching a conductor's top and bottom surfaces.

$$C = \begin{array}{ccc} C_{10} + C_{12} + C_{ij} & -C_{12} & -C_{1j} \\ -C_{21} & C_{20} + C_{21} + C_{2j} & -C_{2j} \\ -C_{i1} & -C_{i2} & C_{i0} + C_{i1} + C_{ij} \end{array} \tag{3.6}$$

The capacitance matrix defined in (3.6) is usually presented more succinctly as in (3.7) [13]:

$$C = \begin{array}{ccc} C_{11} & -C_{12} & -C_{1j} \\ -C_{21} & C_{22} & -C_{2j} \\ -C_{i1} & -C_{i2} & C_{ij} \end{array} \tag{3.7}$$

The capacitance C_{11} is seen to represent the *total capacitance* of the etches: the capacitance to ground of etch 1 (C_{10}) plus all of the mutual capacitance terms when those etches are also grounded. Therefore, all of the terms in a row or column are summed to find the capacitance to ground of a given etch [14]. Incidentally, the same result can be obtained by multiplying (3.6) by a single column matrix filled with ones. The resulting matrix shows C_0 for each conductor.

The capacitance matrix in pF/inch length for the three-conductor system shown in Figure 2.4 appears in (3.8):

$$C = \begin{array}{ccc} 1.706 & -0.342 & -0.0278 \\ -0.342 & 1.789 & -0.342 \\ -0.0278 & -0.342 & 1.706 \end{array} \quad \text{pF/inch} \tag{3.8}$$

The significance of the mutual terms is shown in Example 3.4's calculations.

Example 3.4

Using the matrix in (3.8) and assuming the etches are 1 in in length:
(a) What is the capacitance to ground of etch 2?
(b) What switching capacitance does etch 2 have when etches 1 and 3 remain at 0V?
(c) What is its switching capacitance when all three conductors simultaneously switch at the same rate and with the same final voltage?

Solution

(a) The capacitance of etch 2 is represented by the second column in (3.8). Therefore, the etch capacitance to ground is found by summing the terms in column 2:

$$C_{20} = 1.789 + 2(-0.342) = 1.105 \text{ pF}$$

(b) This situation is illustrated in Figure 3.3. As shown, etch 2's switching capacitance is the sum of C_{20} and all of its mutuals. Because from (3.7) that is also the definition of C_{22}, the switching capacitance of etch 2 is $C_{22} = 1.789$ pF.
(c) This is the situation depicted in Figure 3.4. In this case, the switching capacitance is just the capacitance to ground (C_{20}), calculated in part (a) as 1.105 pF.

The capacitance of etch 2 is seen to vary from between 1.11 to 1.79 pF (a change of over a third) just by the behavior of its neighbors. The ramifications of this switching interaction are described in Chapter 8.

3.6   Dielectric Losses

Applying a voltage across the terminals of a perfect capacitor causes displacement current to flow as the dielectric becomes polarized [15, 16]. This polarization is illustrated in Figure 3.6 and results from the electric field slightly displacing the electrons in their orbits, in effect creating atomic dipoles. Part (a) shows the lattice when no field is applied, while parts (b) and (c) show the effects of an alternating voltage.

Conduction current will also flow because the dielectric is an imperfect insulator and so has some resistivity. This effectively places a conductance in parallel with the perfect capacitor and allows what amounts to a frequency-dependent leakage current to pass between the capacitor's terminals. An etch's total capacitor current is therefore the sum of its displacement and conduction currents.

A circuit model for a capacitor especially suited for transmission line use appears in Figure 3.7. It consists of a perfect capacitor C having no conduction current in parallel with a conductance G, which allows conduction current to flow between the plates. This model is used in Chapter 5 when determining transmission line impedance, while the model presented for discrete capacitors in Chapter 10 (Figures 10.3 and 10.5) represents the loss as a series element [equivalent series resistance (ESR), which as shown in Chapter 10, includes skin effect resistive losses].

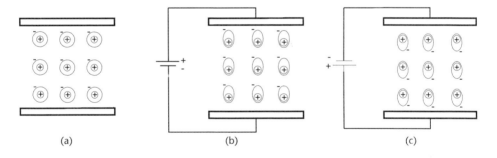

Figure 3.6   Electron displacement with (a) no field, (b) positive voltage, and (c) negative voltage.

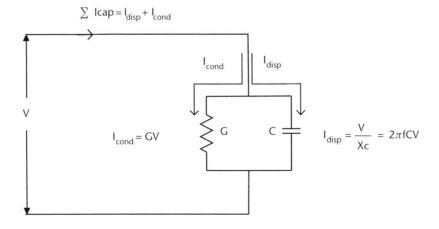

Figure 3.7   Capacitor circuit model.

3.6.1 Reactance and Displacement Current

A capacitor's reactance determines the amount of displacement current that flows at a particular frequency when a capacitor is connected to ac.

If in (3.2) the voltage varies sinusoidally, it's straightforward to show that current and voltage are related as shown in (3.9) [17]:

$$I_d = 2\pi f C V \tag{3.9}$$

where I_d is the displacement current. The quantity $2\pi f C$ has the units of siemens (or, sometimes by the outdated *mhos*, but in either case the reciprocal of ohms). The capacitors reactance is defined as (3.10):

$$X_c = \frac{1}{2\pi f C} \tag{3.10}$$

The reactance (X_c) has ohms as units and the relationship between displacement current and the capacitor's terminal voltage becomes (3.11):

$$I_d = \frac{V}{X_c} \tag{3.11}$$

The implicit understanding in (3.11) is that the capacitor's total current leads the voltage by 90°.

3.6.2 Loss Tangent

As previously discussed, the dielectric used to separate the palates of a capacitor will have a resistivity ρ that allows a conduction current (modeled as a leakage current) to flow in addition to the displacement current considered earlier. Naturally, a capacitor with a high-quality dielectric has small conduction current compared to its displacement current. In fact, the ratio of these two currents is a figure of merit for a dielectrics quality. As shown in (3.12), this ratio is called the *loss tangent* (LT) or *dissipation factor* (Df).

$$\text{Loss Tangent} = Df = \frac{I_{conduction}}{I_{displacement}} \tag{3.12}$$

The curious name *loss tangent* comes about because on a plot of displacement current versus conduction current (an *Argand drawing*, a version of which appears in Figure 3.8), the tangent of the angle δ is the ratio of the conduction current to the displacement current. The angle δ is called the *loss angle*, and this name is sometimes used when specifying losses. However, the tangent of the angle (the *loss tangent*) is more frequently used by PWB laminate suppliers. These terms are interchangeable because for small angles the tangent of an angle expressed in radians is the same as the tangent of that angle. As discussed in Chapter 10, capacitor manufacturers usually prefer to use the more descriptive sounding *dissipation factor* rather than loss angle or loss tangent. All of these terms are interchangeable.

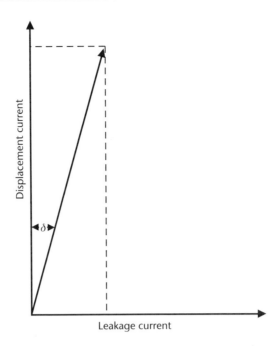

Figure 3.8 Argand drawing showing displacement and conduction currents.

Smaller loss tangents mean a small conduction current–to–displacement current ratio, and thus a more ideal capacitor. This makes intuitive sense: for a good capacitor the current should be dominated by displacement current, not leakage. Typical loss tangent values for ceramic capacitors lie in the 0.001 to 0.05 range [18]. Epoxy resins used as PWB laminates have values in the 0.025 range, but other PWB laminate materials have values under 0.01. Table 3.2 records ε_r/loss tangent at 1 MHz, 1 GHz, and 10 GHz (6 GHz for PPO) as specified by manufacturers' data sheets. These are worst-case manufacturing limits; production values are typically lower and will increase with thickness.

3.6.3 Calculating Loss Tangent and Conductance G

The ways is which dielectric losses affect transmission lines' impedance and propagation characteristics are discussed in Chapters 5 and 7. The conductance G appearing in the model shown in Figure 3.7 is used in transmission line work to represent the etch capacitance loss. It's therefore necessary to have a way to relate G and the loss tangent because material suppliers specify the loss tangent rather than G.

This relationship may be obtained in many ways, with most electromagnetics texts equating the loss tangent with the ratio of the complex to the real portions of the dielectric constant. From this a relationship involving the materials conductivity and dielectric constant is developed (for example, see [34]). While physically sound, this line of attack is not the most instinctive. Instead, we'll use a more intuitive approach that treats the loss as a leakage current and stay with the definition of the loss tangent given in (3.12) [35]. To do so the currents $I_{displacement}$ and $I_{conduction}$ must be found.

Table 3.2 Relative Permittivity/Loss Tangent for Some PWB Laminates

Laminate	1 MHz	1 GHz	10 GHz	Notes/Reference
FR4 Epoxy	3.6/0.032			See [19]
FR4	4.8/0.015	4.3/0.025		Min to max range as reported by several vendors
FR226	4.5/0.019			Terafunctional (140°C Tg); often used as "FR4"[20]
FR370	4.4/0.012			Terafunctional II (170°C Tg); often used as "FR4" [21]
FR408	3.6/0.009	3.5/0.009		FR4 (180°C Tg) [22]
G200		4.1/0.013	3.9/0.009	BT (185°C Tg) [23]
LD-621	3.2/0.004	3.1/0.004	3.1/0.005	Polyphenylene ester (190°C Tg) [24]
MEGTRON	3.8/0.010	3.75/0.011	3.65/0.014	Polyphynoline (PPO) (180°C Tg) 6 GHz (not 10) [25]
N4000-13	3.9/0.009	3.8/0.01	3.6/0.012	Cynate ester with E-Glass (210°C Tg) [26]
N4000-13SI	3.6/0.008	3.5/0.009	3.4/0.01	Cynate ester with enhanced glass; (210°C Tg) [26]
N5000		3.8/~0.014	3.6/0.014	BT (185°C Tg); loss tangent at 2.5GHz = 0.014 [27]
N6000	3.7/0.005	3.4/0.007		Allylated polyphenylene ether (APPE); (210°C Tg) [28]
N6000-SI	3.4/0.003	3.0/0.006		APPE with enhanced glass; (210°C Tg) [28]
N8000	3.8/0.008	3.7/0.011	3.5/0.011	ParkNelco [29] cynate ester (280°C Tg)
ORCER RF-35		3.5/0.0018		Taconic [30] measured at 1.9 GHz; organic-ceramic; (315°C Tg)
P95		4.4/0.016	4.2/0.014	Allied signal [31] polyimide; (260°C Tg)
RO4003			3.4/0.0022	Non-PTFE hydrocarbon-ceramic (280°C Tg) [32]
RO3003			3.0/0.0013	Ceramic filled PTFE (>280°C Tg) [33]
RO3006			6.15/0.0025	Ceramic filled PTFE (280°C Tg); ε_r falls with temperature [33]

It's straightforward to use (3.10) and (3.11) to find $I_{displacement}$, noting that the displacement current calculated will be for the specific frequency used in (3.10). The displacement current is given in (3.12) as:

$$I_{displacement} = \frac{V}{X_c} = 2\pi f C V \qquad (3.13)$$

The conduction current is found from Ohm's law in terms of the conductance G as (3.14):

$$I_{conduction} = \frac{V}{R} + G V \qquad (3.14)$$

For the materials and frequency ranges of interest in this book, $I_{conduction}$ is seen to be the same for all frequencies, and for the frequency ranges shown in Table 3.2, the loss tangent values are also seen to have a fixed value over frequency. This correctly suggests that as (3.14) varies with frequency and (3.13) does not, G will be frequency dependent.

After substituting these into (3.12), some straightforward algebra yields an equation for G in terms of the loss tangent $tan (\delta)$ (3.15):

$$G = 2\pi f C_p \tan(\delta) \tag{3.15}$$

where G is the conductance in siemens, f is the frequency (in hertz), C_p the capacitance value, and $tan(\delta)$ the loss tangent. As expected, higher loss tangent values produce greater values of G, signifying higher conduction current (more loss) at a given frequency. Notice that, as expected, G increases with frequency. This is necessary because the displacement current increases with frequency, forcing a larger conductance G if $tan(\delta)$ is to remain fixed. If C_p is given as a capacitance per length (as will be the case with microstrip and stripline etches), G computed with (3.15) will also be per unit length.

The worked example below shows how to apply (3.15). An intuitive feel for conductance and the significance of the conductance on signal degradation is provided in Chapters 5 and 7.

Example 3.5

Find the conductance value at 250 MHz, 1.25 GHz, and 2.5 GHz for a 6-in long 50-Ω stripline etch having a capacitance of 3.6 pF/inch, assuming the loss tangent is 0.025.

Solution

From (3.15) for a 1-in long line, G = 141μ siemens at 250 MHz, 707μ siemens at 1.25 GHz, and 1.4-m siemens at 2.5 GHz. These values are per inch length of line. Because in this example the line is 6 in long, the final values will be six times larger.

Clearly the dielectric losses as represented by G are increasing with frequency. This example points out the problem many circuit simulators have when computing transmission line losses: the simulation results will be in error unless the simulator can uniquely calculate the transmission line parameters for each of the frequencies contained in a waveform (the waveform's harmonics, as discussed in Chapter 7). In general, time-domain simulators can only do this with difficulty. In fact, failure to apply unique frequency-dependent values for ε_r, G, and conductor resistance for each harmonic contained in a waveform diminishes the usefulness of such a simulation.

3.7   Environmental Effects on Laminate ε_r and Loss Tangent

Many laminates are comprised of an E-glass fiber fabric composed of calcium, aluminum, boron, and silicon saturated with an epoxy resin. Various epoxy resin systems are used to make the laminates, but in general the filaments have a higher dielectric constant (~6.5, see Table 3.1) than the resin. The difference can approach 2:1, depending on the resin system.

The ratio of fibers to the amount of epoxy (the glass-to-resin ratio) determines the laminate's final ε_r value. A thick application of resin will result in a sheet having a dielectric constant closer to the resin than to the E-glass.

Similarly, when manufacturing multilayer boards, fabrication shops often adjust the amount of resin they apply between laminate sheets to meet an overall board thickness specification. In fact, although resin thickness is usually assumed to be the same for each layer, adjusting the thickness by adjusting the resin content results in a different dielectric constant on various layers.

It's worth noting that some mid-range and higher performance laminates employ fibers and resins that have dielectric constants having values similar to each other. As the constituents have nearly the same value, a high resin content layer will have nearly the same ε_r as one having less resin. Said another way, ε_r is less affected by the glass-to-resin ratio in these systems, and so layer-to-layer differences are smaller.

3.7.1 Temperature Effects

For most epoxy-based laminates, ε_r increases and loss tangent falls as temperature rises. However, ε_r decreases in some PTFE composite laminates (such as Rogers RO3006, RO3010, RO5870, and RO5880 [32]). The severity of the change depends on the chemistry, moisture content, and thickness (resin content) of the laminate, but in general ε_r usually increases in epoxy-based laminates by less than 10% over a 25°C to 100°C temperature span [36], regardless of moisture content. However, loss tangent can fall significantly as temperature increases.

The largest loss tangent changes occur when the laminate has low moisture content. When dry, the loss tangent of typical FR4 can decrease by more than 40% over a temperature span of 25°C to 80°C [36].

3.7.2 Moisture Effects

Laminate moisture uptake increases the dielectric's conductivity and thus its leakage. From (3.12), this causes the loss tangent to increase. Again, the severity of the change is dependent on laminate chemistry and thickness, making some laminate types more prone to moisture absorption than others. Polyamide laminates typically have the greatest affinity for water, followed by the epoxy based FR4-class materials. At the other extreme are laminates based on PPO/epoxy and PTFE materials, as these are essentially hydrophobic. Data sheets for these materials show at least an order of magnitude reduction in moisture absorption as compared to FR4-type epoxies (for example, [30, 37]).

After 200 hours of exposure to 90% relative humidity, FR4 epoxy type laminates can experience an increase in loss of up to a 35% and a dielectric constant increase ranging from a few percent to over 10% [25, 36]. Dry laminates have lower loss tangent values but show the largest change with frequency.

Moisture uptake also affects boards' mechanical characteristics. This topic is outside the scope of this book, but here it's noted that water absorption is associated with a lowering of Tg and increasing corrosion and has been known to cause the formation of conductive filaments along the glass filaments [38].

3.8 Summary

The relative permittivity (ε_r, DK, or K) of a dielectric [given in (3.3)] shows how much the capacitance of a structure increases over what it would be with air as the dielectric. For most materials used as PWB laminates, ε_r lies in the 3.5 to 4.8 range but can be less than that for some higher performance laminates (see Table 3.2).

For most laminates ε_r falls less than 10% as frequency increases from 1 MHz to 1 GHz, but some higher performance laminates show significantly less change (see Table 3.2). Because of the general nature of its specification, the ubiquitous FR4 shows the greatest variation with frequency across vendors (exceeding 10%), but in practice a given circuit board fabricated on FR4 will experience less variation.

The loss tangent (also called the loss angle) is a measure of a dielectrics loss as it expresses the ratio of capacitors' conduction current (leakage) to displacement current (3.12). For PWB laminates, losses generally increase with frequency, temperature, and increasing moisture content. The leakage current loss is modeled as a conductance G whose value is frequency dependent.

Laminates fall into one of three loss categories: those having loss tangents over 1% (generally the least expensive laminates); those having losses between 0.5% and 1%; and those with losses under 0.5% (generally the highest cost laminates). Lower loss tangents generally correspond to laminates having lower ε_r and thus lower crosstalk.

Mutual capacitance can be expressed in a capacitance matrix (3.6) and depending on the severity can cause a significant change in an etch's impedance when its neighbors switch. This leads to crosstalk, jitter, and ISI (presented in Chapters 7 and 8).

References

[1] Sibley, M., *Introduction to Electromagnetism*, Chapter 2, London: Arnold Press, 1996.

[2] Miner, G. F., *Lines and Electromagnetic Fields for Engineers*, Oxford: Oxford University Press, 1996, p. 657.

[3] Federal Telephone and Radio Corporation: *Reference Data for Radio Engineers*, 3rd Ed., 1949, pp. 50–52.

[4] Saums, H. L, and W. W. Pendleton, *Materials for Electrical Insulating and Dielectric Functions*, Rochelle Park, NJ: Hayden, 1973.

[5] Shugg, W. T., *Handbook of Electrical and Electronic Insulating Materials*, 2nd Ed., New York: IEEE Press, 1995.

[6] Kaye, G.W.C., et al., *Tables of Physical and Chemical Constants*, 11th Ed., London: Longmans, 1956.

[7] Saint-Gobain Vetrotex Textiles, "E, R and D Glass Properties Textiles Data Sheet," Chambery, France, Q2/2001.

[8] Bogatin, E., "Design Rules for Microstrip Capacitance," *IEEE Trans. CHMT*, Vol. 11, No. 3, September 1988, pp. 253–259.

[9] Hurt, J. C., "A Computer-Aided Design for Hybrid Circuits," *IEEE Trans. CHMT*, Vol. 3, No. 4, December 1980, pp. 525–535.

[10] Nishiyama, H., and M. Nakamura, "Capacitance of Disk Capacitors," *IEEE Trans. CHMT*, Vol. 16, No. 3, May 1993, pp. 360–366.

[11] Miner, G. F., *Lines and Electromagnetic Fields for Engineers,* Oxford: Oxford University Press, 1996, p. 461.

[12] Paul, C. R., *Analysis of Multiconductor Transmission Lines,* New York: John Wiley & Sons, 1994, p. 55.

[13] Bakoglu, H. B., *Circuits, Interconnections and Packaging for VLSI,* Reading, MA: Addison Wesley, 1990, p. 294.

[14] Young, B., *Digital Signal Integrity,* Englewood Cliffs, NJ: Prentice Hall, 2001, p. 235.

[15] Kraus, J., and K. Carver, *Electromagnetics,* 2nd Ed., New York: McGraw Hill, 1973.

[16] Sibley, M., *Introduction to Electromagnetism,* London: Arnold Press, 1996, pp. 150–157.

[17] Skilling, H. H., *Electrical Engineering Circuits,* 2nd Ed., New York: John Wiley and Sons, 1965, p. 36.

[18] ANSI/EIA 198-1-E-97, "Ceramic Dielectric Capacitors, Class I, II, III, IV, Part I: Characteristics and Requirements," Electronics Industries Alliance.

[19] ASM International, *Electronic Materials Handbook, Vol. 1: Packaging and Materials,* Park, OH, 1979.

[20] Polyclad Laminates, "PCL-FR-226 Product Data Sheet," Franklin, NH, April 2003.

[21] Polyclad Laminates, "PCL-FR-370 Product Data Sheet," Franklin, NH, April 2003.

[22] Isola Laminate Systems, Inc., "FR408 Product Data Sheet," No. 5035/9/99, LaCrosse, WI, 1999.

[23] Isola Laminate Systems, Inc., "G200 Product Data Sheet," No. 5027/2/02, Chandler, AZ, 2002.

[24] Polyclad, Cookson Electronics PWB Materials & Chemistry, "PCL-LD-621 Data Sheet," Franklin, NH, 2002.

[25] Matsushita Electronic Materials, Inc., "Megtron R5715 Technical Brochure," Number 199707-1.5Y, Forest Grove, OR, 1997.

[26] Nelco Park Products, "N4000-13SI Data Sheet," Rev. G5-03, Fullerton CA.

[27] Nelco Park Products, "N5000 Data Sheet," Rev. E2-03, Fullerton CA.

[28] Nelco Park Products, "N6000-SI Data Sheet," Rev. E2-02, Fullerton CA.

[29] Nelco Park Products, "N8000 Data Sheet," Rev. D2-03, Fullerton CA.

[30] Taconic Advance Dielectric Division, "ORCER RF-35 Data Sheet," Petersburg, NY.

[31] Isola Laminate Systems, Inc., "P95 Product Data Sheet," No. 5028/2/02, Chandler, AZ, 2002.

[32] Rogers Corporation, Microwave Materials Division, "R04000 Series High Frequency Circuit Materials," Revision 4/00, Chandler, AZ, 2000.

[33] Rogers Corporation, Microwave Materials Division, "R03000 Series High Frequency Circuit Materials," Revision 10/98 Chandler, AZ, 2000; "High Frequency Laminates Data Sheets," Chandler, AZ, 1998.

[34] Miner, G. F., *Lines and Electromagnetic Fields for Engineers,* Oxford: Oxford University Press, 1996, p. 656.

[35] Sibley, M., *Introduction to Electromagnetism,* Chapters 4 and 6, London: Arnold Press, 1996.

[36] Khan, S., "Comparison of the Dielectric Constant and Dissipation Factors of Non-Woven Aramid/FR4 and Glass/FR4 Laminates," Technical Note, DuPont Advanced Fibers Systems Division, Richmond, VA.

[37] Rogers Corporation, Microwave Materials Division, "High Frequency Laminates Data Sheets," Chandler, AZ, 1999.

[38] Pecht, M. G., et al., "Moisture Ingress Into Organic Laminates," *IEEE Trans. Components and Packaging Tech.,* Vol. 22, No. 1, March 1999, pp. 104–110.

CHAPTER 4

Inductance of Etched Conductors

4.1 Introduction

Generally, high-speed circuit designers study inductors as lumped elements such as coils or chokes. While studying inductors as discrete devices provides circuit-level insight, it misses the actual nature of inductance, an understanding vital when signaling at high speeds. On the other hand, using electromagnetics to study the fundamental, underlying nature of inductance often has little practical value to the design engineer, as this approach generally focuses on the sometimes difficult mathematics rather than the development of intuitive insight.

This chapter takes a hybrid electromagnetic field and linear circuit approach to the study of inductance (especially the concept of *loop inductance*). A concise review of key electromagnetic principals is initially presented, but the remainder of the chapter focuses on applying familiar linear circuit and network theory concepts. Those interested in the mathematics of the underlying physics (as well as for many of the proofs not offered here) are referred to [1–3]. A more elementary approach to the mathematics is taken by [4, 5].

Consistent with the custom throughout this book the examples are calculated in *engineering units* (inches rather than meters), but all physical constants are provided both in metric and engineering units. The reader may recalculate the examples with metric units if desired.

4.2 Field Theory

We begin the discussion of inductance with the briefest review of field theory concepts. The focus will turn to circuit theory once these underlaying principals have been introduced.

Current flowing in a line causes rings of magnetic field lines to be created along the lines length perpendicular to (coaxially to) the direction of the current flow. This is illustrated in Figure 4.1. Gausses's law for magnetic fields requires the field lines to form the closed loops shown. The number of magnetic lines per unit area is the *magnetic flux density*, (B, with units of Tesla). The *magnetic field intensity* is symbolized by H, with units of amp-turns/meter. That is, H equals the total current divided by the average magnetic path length.

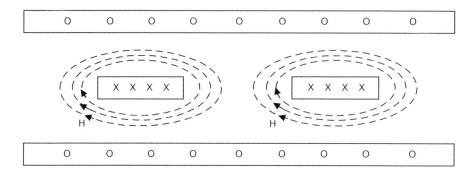

Figure 4.1   Stripline magnetic field lines. Crosses show current flowing onto page, dots show current flowing out of page. Trace separation is large enough to prevent interaction of field lines.

4.2.1   Permeability

The intensity of the magnetic field is related to the corresponding flux density as in (4.1):

$$B = \mu_0 \mu_r H = \frac{\Phi}{A} \qquad (4.1)$$

where μ_0 is the *permeability of free space* (often simply called the permeability and is exactly equal to $4\pi \times 10^{-7}$ H/m, or about 31.919 nH/inch), and Φ is the magnetic flux (with units of *webers*). The area enclosed by the magnetic flux is A.

Some materials are ferromagnetic and have permeabilities greater than μ_0. The ratio of that value to μ_0 is defined as the materials *relative permeability* (μ_r). This is conceptionally similar to a capacitor's relative permativitty ε_r, but for copper (and in fact for most metals used on PWB) $\mu_r = 1$.

In some texts, the ($\mu_0 \mu_r$) product is simply called μ, with no subscript. This ambiguous notation is avoided here. Instead, $\mu_0 \mu_r$ are called out explicitly. Note that because most materials used in PWB work are nonferric μ_r will generally be 1.0 and the ($\mu_0 \mu_r$) product simply reduces to μ_0.

4.2.2   Inductance

From a field's perspective, inductance relates the number of magnetic flux lines in a given enclosed region to the current (I) required to produce the field lines in that region (4.2). Using the definition for Φ given in (4.1) yields (4.2) as a general definition of inductance L:

$$L = \frac{\Phi}{I} = \frac{BA}{I} \qquad (4.2)$$

The significance of (4.2) is that it shows inductance as a property of current flow and the magnetic field contained within a region. Implied in (4.2) is that the round trip path taken (the *loop area, A*) by the current is fundamental when computing inductance.

For there to be current flow in the wire of Figure 4.1, there must also be a path for the current to flow back to the source. If this return path is close by, the loop area is small, making the inductance small. If the return is moved away, the current and number of flux lines remains unchanged, but the loop area containing them is larger. This increases Φ, and thus, from (4.2), the inductance.

Depicted in Figure 4.1 is the case when current is flowing in two conductors referenced to a common return. It's clear that the magnetic field lines extend beyond each conductor, and they could easily impinge on a neighboring conductor. This magnetic coupling (called *mutual inductance*) induces a voltage on the neighboring conductor and results in crosstalk. However, as discussed in Section 4.5, mutual inductance can be used to advantage to lower the total circuit inductance.

4.2.3   Internal and External Inductance

Figure 4.1 implies the magnetic field lines are only external to the conductor, but at low frequency fields also exist within the conductor. These fields give rise to an internal inductance just as the fields surrounding the conductor produce an external one. In fact, the total inductance of a wire is the sum of these two inductances. At frequencies high enough for the skin effect to be well developed, the internal inductance approaches zero due to the migration of current away from the conductor's interior. Consequently, for practical purposes a wire's high frequency inductance is simply its external inductance. External inductance is a property of a conductor's geometry and relationship to its return path—and is what field solvers report.

It can be shown [6] that the internal self inductance for a round wire of any diameter is exactly 50 nH per meter length (i.e., about 1.27 nH per inch).

4.2.4   Partial Inductance

From (4.2), inductance is the proportionality factor relating current flow to the magnetic flux in a region. Of course, for current to flow it must make its way from the source to the load and back. It's the area of this closed loop that determines the circuit's inductance. Estimating the loop area (and thus, by definition, inductance) for complex shapes (such as formed pins in a connector) can be difficult, and computation of the total loop traversed by a long, straight trace is tedious and unnecessary, as each segment of the trace is identical and evidently contributes equal amounts to the total inductance.

The concept of *partial inductance* [7, 8] addresses this. The concept is that a complex shape may be segmented into many smaller, regular pieces, each with its own inductance. The shape's total inductance is the sum of these partial inductances. Mutual inductance can be included to model inductive coupling between segments and to other segments. For a model of partial inductors to be a proper representation of the structure's total inductance, a common reference point must be assumed for all of the segments. In fact, the noise voltage developed by the sum of these partial inductances can only be properly measured with respect to this one common point (often called the model's *return* or *reference*). It is critical not to lose sight of this when using computer models (such as SPICE) to predict inductive circuit behavior.

Notice that (4.1) does not include ε_r. In fact, the surrounding material's dielectric constant is not a factor in determining the inductance. As is shown in Section 4.6, it's the trace's geometry, not the background dielectric, that matters. This means that if a trace's thickness, width, and spacing from the return planes remains the same but ε_r is changed (as would happen if the same PWB layout were used with different laminate systems), the traces capacitance would change but its inductance would not. As described in Chapter 5, changing capacitance without a corresponding change in inductance alters the transmission line's impedance and delay characteristics.

Developing package and PWB partial inductance models is beyond the scope of this book. See [9] for an excellent discussion of these techniques and on ways the reference point can be altered by manipulating matrices of partial inductors.

4.2.5 Reciprocity Principal and Transverse Electromagnetic Mode

Lines without too much loss that have the same dielectric everywhere (such as stripline) are capable of propagating energy such that the electric and magnetic fields are transverse (i.e., crosswise) to the direction of energy flow. Such a condition is called *transverse electromagnetic* (TEM) mode of propagation.

A detailed discussion of TEM is outside the scope of this book. References [1, 2, 6] provide an in-depth analysis. Here it's noted that TEM requires two or more conductors, with at least one acting as the return for the other(s). For this reason, TEM is sometimes called the *transmission line mode of propagation*. With TEM, an analogy exists between the electric and magnetic fields leading to the familiar capacitor/inductor circuit transmission line model discussed in Chapter 5. In this circuit model, the electric field component is represented by a capacitor, and the magnetic field, by an inductor. This analogy extends further and allows for simple propagation, impedance, and loss calculations. Although other modes are possible, it's desirable for TEM to be the dominant mode of propagation on PWB trace. For this to be the case, resistive losses must be low and the frequency must be high enough so that the internal inductance approaches zero. However, other modes will be set up if the propagating wavelength becomes a significant fraction of the conductor's physical dimensions such as thickness, spacing to the return path, and width. It can be shown [10] that for a parallel plate transmission line, TEM will be the only mode of propagation if the line dimensions are electrically small. That is, TEM will be the only mode propagating if the line's physical dimensions are a small fraction of a wavelength of the highest frequency component being transmitted. The relationship between wavelength, frequency, and dielectric constant is presented in Chapter 5. For now, we note that the wavelength is about 0.56 in for stripline on FR4 at 10 GHz. Only TEM will propagate, provided the lines physical dimensions are much smaller than this (i.e., 50 mils). This increases to nearly 5.5 inches at 100 MHz.

In fact, the trace width, spacing, and layer thickness of typical digital PWB's are smaller than these minimum dimensions. This means that TEM (or quasi-TEM, as in microstrip where the dielectric is nearly homogeneous) will be the mode of propagation on PWB transmission lines at least up to these frequencies, and is the mode tacitly assumed throughout this book. In the case of TEM propagation, the relationship between a transmission line's external inductance and capacitance is given in (4.3):

$$LC = \mu_0 \mu_r \varepsilon_0 \varepsilon_r \qquad (4.3)$$

The relationship in (4.3) is often called the *reciprocity principal* [6] and allows for inductance calculation at high frequency (where the skin effect is fully developed) if the capacitance of a structure is known (or can be measured accurately). Conversely, the capacitance may be calculated if the inductance is known. It's only strictly applicable in TEM situations where the dielectric is homogeneous, such as in stripline.

Example 4.1

A 5-mil-wide, 60-Ω copper stripline fabricated on a laminate having $\varepsilon_r = 4.0$ is measured at 100 MHz to have a capacitance of 2.9 pF/in length. What is its partial self inductance?

Solution

Because the frequency is high enough for the skin effect to be fully developed (see Chapter 2), the internal inductance approaches 0, leaving only the external inductance. As the conductor is copper, $\mu_r = 1$. From (4.3), and using inch-based units for μ_0 and ε_0, the inductance is:

$$L_{ext} = \frac{\mu_0 \varepsilon_0 \varepsilon_r}{C} = \frac{31.919\ \text{nH} \times 224.79\ \text{fF} \times 4}{2.9\ \text{pF}} = 9.9\ \text{nH}$$

The reciprocity principal is a valuable tool in checking the results from field solvers, especially when used to determine the capacitance matrix from the inductance matrices as described in Section 4.4.

4.3 Circuit Behavior of Inductance

The previous section used electromagnetic concepts from field theory to describe the creation of inductance, with no regard to its effects on circuit operation. In this and subsequent sections, inductance is regarded as a circuit element. To illustrate inductive effects, consider a 5-mil-wide, 1-in-long, 50-Ω microstrip terminated in a 50-Ω load, as shown in Figure 4.2. For purposes of this discussion, the trace will have no capacitance or resistance.

A circuit schematic of the Figure 4.2 setup appears in Figure 4.3. From Ohm's law, the ammeter should show a 20-mA step response when the switch is closed.

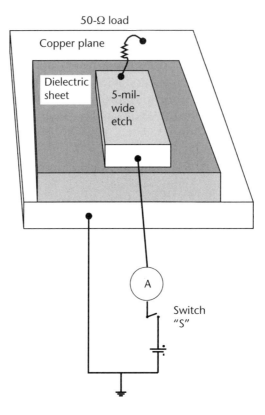

Figure 4.2 Microstrip connecting load to source.

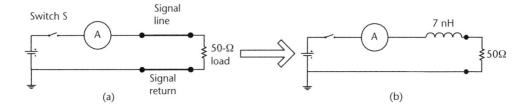

Figure 4.3 Schematic of Figure 4.2 topology, showing signal and return inductance may be represented by one 7-nH inductor, provided the proper reference point is preserved. In (a) the load is connected to S by inductance in the signal and return lines. These have been combined in (b) to a single inductor.

Instead, the ammeter shows the current rising exponentially, requiring nearly 1 ns to fully reach 20 mA. This is shown in Figure 4.4.

The reluctance of the current to change suddenly is due to the inductance of the path connecting S to the load. It's often said that a trace or wire has a certain amount of inductance, but it's actually the source/return loop that possesses this property. By definition [see (4.2)] it's not possible to determine the signal's inductance unless the return path is known. In fact, the waveform in Figure 4.4 is only correct with respect to the reference connection shown. Inductance is therefore a property of the

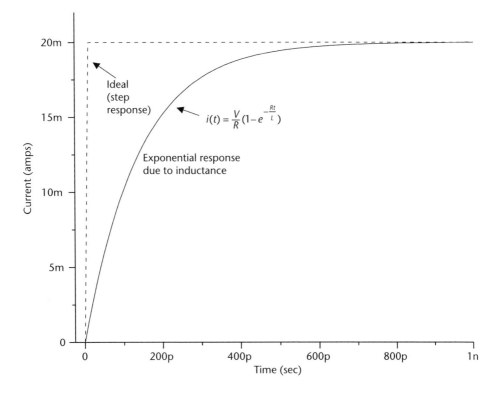

Figure 4.4   Ideal and actual current response of the inductive circuit illustrated in Figure 4.3.

physical relationship between a signal and its return. As shown in Figure 4.3(b), on schematics it's often convenient to lump the inductance totally in the signal lead. If done carefully, this gives the proper circuit response. However, when doing so it's important not to lose sight that inductance is a property of the current flowing in an area [as described by (4.2)], and the relationship to the return path is key.

4.3.1   Inductive Voltage Drop

Faraday showed that a changing electromagnetic field causes an opposing voltage to be developed along the wire in proportion to the rate of change of the current flowing through that wire (4.4):

$$e = -L\frac{di}{dt} \tag{4.4}$$

where the inductance L (units henrys) is the proportionality factor relating the elapsed time (dt) it takes the current to change by di amps.

Example 4.2

As shown in Figure 4.5, an application-specific integrated circuit (ASIC) simultaneously drives four 50-Ω loads with a 1-ns rise time through a 5-nH power lead inductance to a perfect 2.5-V source. What is the power supply noise voltage, assuming no lead resistance or decoupling capacitance?

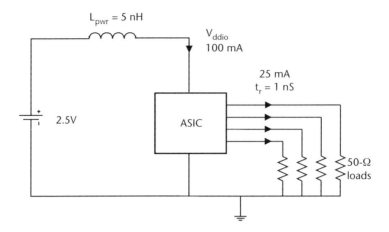

Figure 4.5 ASIC driving four 50-Ω loads as described in Example 4.2.

Solution

From (4.4), the voltage across L_{pwr} is: $e = -5 \text{ nH} \dfrac{4 \times 25 \text{ mA}}{1 \text{ ns}} = -0.5\text{V}$, with the nega-

tive sign implying a voltage drop. Therefore, during the switching event, V_{ddio} will fall from 2.5V to 2.0V.

Of course, in practice the leads will have some resistance, and the voltage drop due to the resistance will add to the voltage computed with (4.4). In general, the voltage drop along the loop is the sum of the resistive and inductive drops (4.5):

$$v = iR + L\frac{di}{dt} \tag{4.5}$$

If in Example 4.2 the lead resistance is 50 milliohms, the iR drop will be 5 mV. This is an insignificant fraction of the 500-mV instantaneous drop due to the inductor.

4.3.2 Inductive Reactance

With some simple algebra, (4.4) can be manipulated to determine X_l, the inductor's *reactance* (4.6), assuming current varies sinusoidally [11]:

$$X_l = 2\pi f L \tag{4.6}$$

The reactance (X_l) has ohms as units, and the relationship between the current through the inductor and its terminal voltage becomes (4.7):

$$I = \frac{V}{X_l} \tag{4.7}$$

Tacitly assumed in (4.7) is that the current lags the voltage by 90°. This is just the opposite of a capacitor, where the current leads the voltage by 90°.

4.4 Inductance Matrix

As is the case with resistance and capacitance, a matrix can be used to show the relationship between a trace's self and mutual inductance. The *inductance matrix* (4.8) shows the inductance of a trace with respect to its return (the self inductance) on the main diagonal, with the currents in all other conductors zero [10]. The off diagonal terms show the coupling inductance appearing between traces (the *mutual inductance*).

$$L = \begin{matrix} L_{11} & L_{12} & L_{13} \\ L_{21} & L_{22} & L_{23} \\ L_{31} & L_{32} & L_{33} \end{matrix} \qquad (4.8)$$

For example, the self inductance of conductor 1 is L_{11}, and the inductance coupling traces 1 and 2 is L_{12}.

4.4.1 Using the Reciprocity Principle to Obtain the Inductance Matrix from a Capacitance Matrix

Example 4.1 demonstrated the use of the reciprocity principle to obtain a conductor's inductance from its capacitance. In the same way, it's possible to obtain an inductance matrix from a capacitance matrix, as demonstrated in Example 4.3.

Example 4.3

Assuming stripline and that $\varepsilon_r = 4.0$, compute the inductance matrix from the capacitance matrix (3.8) given in Example 3.4.

Solution

As was the case for Example 4.1, this is a direct application of the reciprocity principal (Section 4.2.5) by solving (4.3) for the inductance L, but in this case it's necessary to use matrix algebra to solve for the inductance, as the capacitance is given in matrix form in (3.8). Although tedious by hand (for example, see [12]), this is easily done using computational software (such as Mathcad [13] or MATLAB [14]). Keeping the mutual capacitance terms negative in (3.8), assuming $\varepsilon_r = 4.0$, and applying the equation for L developed in Example 4.1 yields the following inductance matrix:

$$L = \begin{matrix} 17.55 & 3.55 & 0.997 \\ 3.55 & 17.398 & 3.55 \\ 0.997 & 3.55 & 17.55 \end{matrix} \quad nH/in \qquad (4.9)$$

4.5 Mutual Inductance

Equation (4.4) shows how inductive coupling can cause voltages to be induced between wires. If L in (4.4) is an inductance appearing between an aggressor and a victim wire, then a changing current flowing in the aggressor will induce a voltage in the victim proportional to the rate of change of the aggressor's current. The proportionality factor is called the *mutual inductance*, with units of Henrys. The general symbol for mutual inductance is L_m (also called M). As shown in (4.18), array notation may be used to specify the coupling between specific conductors. The key roll mutual inductance plays in developing crosstalk voltages between signals is discussed in Chapter 8.

Although self inductance must always be positive, mutual inductance may have either a positive or negative value. The choice depends on whether the coupling results in the voltage having the same or opposite polarity from the aggressor voltage. As shown in Example 4.4, this can be used advantageously to lower the loop inductance between a signal and its return or of a power and its return line.

When used to advantage, mutual inductance will beneficially reduce the total loop inductance in the case where inductors are effectively in series. However, parallel inductors generally do not benefit in the same way, and the inductance of multiple inductors in parallel will not be reduced as strongly as expected. These two situations are described in Sections 4.5.2 and 4.5.3.

4.5.1 Coupling Coefficient

An alternative way to indicate inductive coupling between circuits is to use a unitless ratio called the *coupling factor* or the *coupling coefficient* (symbol k) (4.17):

$$k = \frac{L_{12}}{\sqrt{L_1 L_2}} = \frac{M}{\sqrt{L_1 L_2}} \tag{4.10}$$

where L_{12} is the mutual inductance between inductors L_1 and L_2. The coupling coefficient lies between ± 1, with values further from zero signifying circuits that are more tightly coupled. Many circuit simulators (most notably SPICE) specify mutual inductance with k rather than L_m (i.e., L_{12}). In this case, (4.10) may be used to translate from mutual inductance to coupling coefficient.

The following example shows how to use the inductance matrix to determine the self inductance, the mutual inductance, and the corresponding coupling coefficient.

Example 4.4

Using the matrix in (4.9) and assuming the traces are 1 inch in length:
(a) What is the self inductance of traces 1 and 2?
(b) What is the mutual inductance between traces (1 and 2) and (1 and 3)?
(c) What is the coupling coefficient between traces (1 and 2) and (1 and 3)?

Solution

(a) By definition, the traces' self inductance can be obtained from the inductance matrix by inspection: $L_{11} = 17.55$ nH and $L_{22} = 17.398$ nH.
(b) By inspection, traces 1 and 2 are seen to be coupled by an inductance L_{12} of 3.55 nH. Because traces 1 and 3 are spaced further apart, they are only coupled by $L_{13} = 0.997$ nH.
(c) From (4.10), $k_{12} = \dfrac{3.55}{\sqrt{17.55 \times 17.398}} = 0.203$, and similarly $k_{13} = 0.057$.

4.5.2   Beneficial Effects of Mutual Inductance

Figure 4.6 shows a two-wire system connecting a battery through a switch to a load. The wires each have an inductance with respect to a reference (*ground*), and there are mutual inductances between them. The second part of the figure shows a short-hand way of illustrating the mutual inductance, with the dots showing the phase relationship between L_1 and L_2 and a curved line representing the mutual inductance. As configured, equal and opposite current flows in the two leads. The final portion shows how the network may be represented by a single inductance.

To derive an equivalent single inductor, we note that from (4.4) the voltage in an inductive circuit is proportional to the rate that the current through the total inductance changes times the total inductance. It follows that the total inductance may be found by summing the induced voltage drops and solving (4.4) for the inductance.

In Figure 4.6, mutual inductance L_{12} will cause the current flowing through L_1 to induce a voltage on L_2 that is the opposite of that across L_2. Similarly, the current flowing through L_2 will induce a voltage on L_1 that is in opposition to L_1's voltage. As this is a series circuit, the same current flows in L_1 as in L_2, and because the same mutual inductance appears between L_1 and L_2 as between L_2 and L_1, the total loop inductance is (4.11) [15]:

$$L_t = L_1 + L_2 - L_{12} - L_{21} = L_1 + L_2 - 2L_{12} \tag{4.11}$$

Notice that if the two wires are spaced far apart such that $L_{12} \sim 0$, the power/return loop appears as two independent inductors in series, and, as expected, the total inductance is simply the sum of L_1 and L_2.

However, as the lines are brought closer together, L_{12} is no longer negligible and the total inductance is less than the sum of L_1 and L_2. As illustrated in the following example, this beneficial effect is useful when pinning out connectors, micro-packages, or cabling.

Example 4.5

An ASIC residing on a daughter card signals back to the motherboard through a 3-pin connector. The signal may either be placed between the power and ground pins, as shown in Figure 4.7(a), or the power/ground pins may be adjacent and the signal offset as in Figure 4.7(b). Compute the signal loop inductances for each configuration using the self and mutual inductances shown.

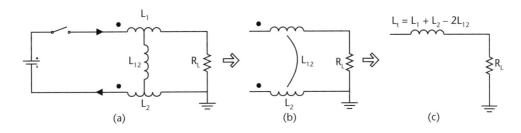

Figure 4.6   Alternate forms for showing mutual inductance L_{12}. Complete circuit is shown in (a); (b) shows shorthand representation; and circuit is reduced to a single inductor in (c).

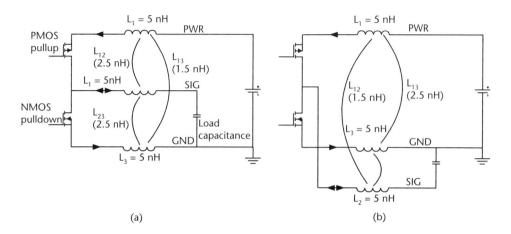

Figure 4.7 Circuit arraignment for Example 4.5 showing options for signaling across 3-pin connector. In (a), signal is centered between power/ground pins; in (b), power/ground pins are adjacent.

Solution

We first observe that in Figure 4.7(a) the signal pin is symmetrically located between the power and ground pins, while in Figure 4.7(b) the signal pin is physically closer to the ground pin.

In Figure 4.7(a), the mutual inductance (and thus the signal inductance) will therefore be the same for a signal switching high or low. However, in Figure 4.7(b), the mutual inductance between signal and ground is lower than the inductance from signal to power. Therefore, when switching low, the signal will have a lower inductance than when switching high. From (4.4), this change in inductance will result in noise voltages (and, evidently, delays) that will differ depending on the switching direction. Also, because impedance is a function of inductance (see Chapter 5), the impedance through the connector seen by the ASIC's driver (if the edge rate is fast enough) will depend on whether the signal is switching high or low. These asymmetrical signal characteristics may not be acceptable in some applications.

A second observation is that mutual inductance between power and ground is lower in Figure 4.7(b) than in Figure 4.7(a). Configuration Figure 4.7(b) will therefore have lower power/ground loop inductance, and thus lower power supply noise.

Computing the loop inductance is a straightforward application of (4.15). The results are tabulated in Table 4.1.

The choice between symmetrical switching inductance at the expense of higher power/ground loop inductance in Figure 4.7(a) verses lower power/ground loop inductance that results in unequal switching inductance of configuration Figure 4.7(b) is a common dilemma that must be decided on a case-by-case basis. Often a carefully considered decoupling scheme can make up for lower mutual coupling (and thus higher loop inductance) between power and ground leads when configuration Figure 4.7(a) is chosen, but little can be done to correct the unequal switching noise if configuration Figure 4.7(b) is selected.

Table 4.1 Loop Inductance Results for Example 4.5

Loop Name	Inductance	Inductance
	Case A	Case B
PWR/SIG	5 nH	7 nH
GND/SIG	5 nH	5 nH
PWR/GND	7 nH	5 nH

4.5.3 Deleterious Effects of Mutual Inductance

Mutual inductance plays a detrimental role in reducing the total loop inductance when two closely spaced wires are put in parallel to carry the same signal (often multiple power or ground connections in cable, or when connector pins or vias are "doubled up" to carry more current). The expectation is that being in parallel, the inductors will behave as *resistors in parallel* and so will have lower total inductance equal to the product over the sum of the two. But this is only true when the inductors are very loosely coupled. From [15], the equivalent inductance of two inductors in parallel is given in (4.12):

$$L_p = \frac{L_1 L_2 - L_{12}^2}{L_1 + L_2 - 2L_{12}} \tag{4.12}$$

If the mutual inductance is negligible, (4.12) does reduce to the familiar product over the sum equation, and the total inductance is indeed the parallel combination of L_1 and L_2. Two identical inductors in parallel would therefore have half the inductance of just one. However, (4.12) shows that the inductance will be higher when the mutual inductance is not zero.

For example, placing an identical wire in parallel with the power lead in case (b) of Example 4.5 lowers the power lead inductance from 5 nH to 3.75 nH. This is 1.25 nH higher than that calculated by the simple product over the sum equation.

The loop equations for inductors in series and parallel are summarized in Figure 4.8.

It's apparent from (4.11) and (4.12) that care must be taken when attempting to lower inductance by using multiple vias to connect together separate power planes, or when multiple pins are used in a connector to improve current carrying capacity. Making a connection by doubling up vias (or pins) will increase the current handling capacity, but from (4.12) the inductance will be something higher than half the inductance of a single via or pin. To get half the inductance , the vias must be placed far enough apart so that the mutual inductance is negligible. On the other hand, placing power and ground vias close to one another (as can be done with a decoupling capacitor) increases the mutual inductance. In this case, this is beneficial because with the current flowing in opposite directions, the overall loop inductance is reduced. This is discussed in Chapter 10.

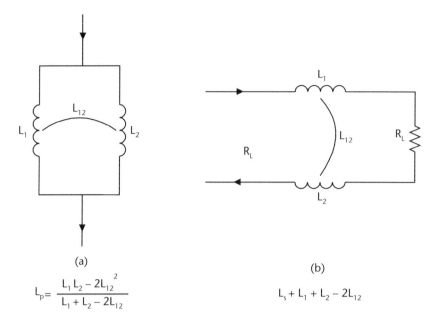

$$L_p = \frac{L_1 L_2 - 2L_{12}^{\;2}}{L_1 + L_2 - 2L_{12}}$$

$$L_s + L_1 + L_2 - 2L_{12}$$

Figure 4.8 Equivalent inductance for inductors in (a) series, and (b) parallel.

4.6 Hand Calculations for Inductance

In general, field-solving software is the best way to determine partial inductance of a conductor, especially in the presence of other conductors. Alternatively, if the capacitance is known (or can be computed) and if the structure supports TEM propagation, the reciprocity principal (Section 4.2.5) may be used to find the partial self and mutual inductance of any shaped conductor or conductor system.

Formulas are available to estimate the inductance of various arraignments and shapes of conductors, with those for round wires being the simplest and most accurate to hand calculate. Results of calculations for nonround conductors such as PWB trace are only approximate because of complications arising from the edge and proximity effects described in Chapter 3. Section 4.6.4 and 4.6.5 presents calculations for microstrip and stripline with those limitations in mind.

4.6.1 Inductance of a Wire Above a Return Plane

Neglecting the internal inductance, the partial-inductance solution for a wire above a return plane is given in [16] as (4.13):

$$L_{ext} = \frac{\mu_0 \mu_r}{2\pi} \ln\left(\frac{2h}{r}\right) \tag{4.13}$$

where r is the wires radius and h the conductor's height above the return plane, measured from the center of the conductor [see Figure 4.9(a)]. Equation (4.13) assumes a perfectly conducting, infinitely wide return plane. In most cases of interest in PWB work, the metal is nonferrous, making $\mu_r = 1$.

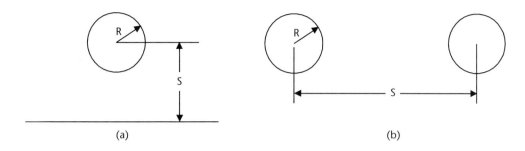

Figure 4.9   Topology for (a) (4.13), and (b) (4.14).

At low frequencies, below where skin effect comes into play and thus the internal inductance is still significant, add 50 nH per meter length (i.e., 1.27 nH/in).

4.6.2   Inductance of Side-by-Side Wires

Neglecting the internal inductance, the partial-inductance solution for the total loop inductance of two side-by-side wires, one acting as a return, is given in [17] as (4.14):

$$L_{ext} = \frac{\mu_0 \mu_r}{\pi} \ln\left(\frac{s-r}{r}\right) \tag{4.14}$$

where r is the wire's radius and s the conductor center-to-center spacing [see Figure 4.9(b)]. Equation (4.14) assumes identical radius for each wire and is useful when computing the loop inductance of a power/ground wire pair, as might appear in a cable.

At low frequencies where the internal inductance is still significant, add 100 nH per meter length (i.e., 2.54 nH/in), as there are two wires creating the circuit.

4.6.3   Inductance of Parallel Plates

The external inductance (L_{ext}) of a parallel plate inductor is given in [1] as (4.15) as:

$$L_{ext} = \mu_0 \mu_r \frac{d}{w} l \tag{4.15}$$

where d is the spacing between the two plates, w the plate width, and l the plate length. As mentioned previously, for nonferrous metals $\mu_r = 1$. Equation (4.10) assumes $w \gg d$ so that [similar to the restrictions placed on the parallel plate capacitor equation of (3.4)] end effects may be ignored. In fact, a great deal of similarity exists between the capacitance equation of (3.4) and the inductance equation of (4.15). In (3.4), the capacitance is seen to increase with the dielectric's permittivity times the ratio of the plate's width to spacing. From (4.15), the inductance is seen to increase with the permeability times the ratio of the plate's spacing to width. Said another way, increasing the plate spacing increases the inductance but decreases the capacitance between parallel plates. From this discussion, wide, closely spaced power/ground planes will have lower inductance than narrow, widely separated plates. Example 4.6 illustrates this.

Example 4.6

What is the inductance of the power/ground system described in Example 3.2?

Solution

To solve this problem, it's first necessary to decide the direction of current flow because [from (4.15)] inductance increases with increasing length but decreases with larger widths. Therefore, interchanging length and width will result in different inductance values. This is in contrast with capacitance, where it's the area (the product of length and width) that matters. In fact, for capacitance the definition of *length* and *width* is inconsequential but is fundamental when calculating inductance (and incidentally, resistance).

Figure 4.10 shows the two possibilities for the 6 × 9 in power/ground plane in Example 3.2. The current is shown flowing along the long dimension in Figure 4.10(a) and along the short dimension in Figure 4.10(b). From Example 3.2, the plate spacing is 5 mils (0.005 in).

Using (4.15) with engineering units makes $\mu_0 = 31.919$ nH/in, and recognizing that for copper $\mu_r = 1$, the inductance of Figure 4.10(a) is:

$$L_{ext} = \mu_0 \frac{d}{w} l = 31.92 \text{ nH} \frac{0.005}{6} 9 = 239.4 \text{ pH}$$

The inductance is reduced to 106.4 pH when the current flows along the short dimension as in Figure 4.10(b).

Configuration Figure 4.10(a) has a loop area of 0.045 in^2 and an inductance of 26.6 pH/in length, while Figure 4.10(b) has a loop area of 0.030 in^2 and an inductance of 17.7 pH/in length. Changing the orientation has reduced the inductance by a factor of 1.5 times.

Example 4.6 gives some insight into the placement of components to reduce overall inductance. For example, locating a voltage regulator at the far corner of a

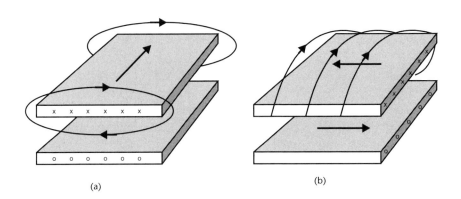

(a)

(b)

Figure 4.10   Magnetic field lines for current flowing along (a) long axis, and (b) short axis in Example 4.6.

PWB often opens up routing channels for signals elsewhere on the board, but this placement will create the highest power supply inductance if it maximizes the area over which the current must travel. Conversely, placing the regulator at the board's center may lower the overall power system loop area but in so doing might adversely constrain signal routing. As is often the case when considering inductive effects, the trade-off between lowering loop inductance and improving routing density is often in conflict and must be made on a case-by-case basis. Trade-offs of signal length, board stackup (especially thickness), and power supply decoupling are hard to generalize, as each application will have its own unique requirements.

4.6.4 Inductance of Microstrip

An equation to compute the partial self inductance of a microstrip that somewhat accounts for fringing is given in [18]. It's recast in a more convenient form in (4.16) and may be used for single or multiple microstrips sitting side by side.

$$L_{sms} = K_{ms} \ln\left(2\pi \frac{h}{w}\right) \qquad (4.16)$$

where K_{ms} = 200 nH/m or about 5.08 nH/in.

Because there are no correction factors for proximity effect, (4.16) is most accurate when computing the partial self inductance of a single trace. In that case, the error is within 5% as compared to field-solving software but rises to nearly 25% when other conductors are close by. As expected, greater separation results in lower errors. An alternative to (4.16) is to backcompute the inductance (or the capacitance) once the impedance and time of flight are known. Chapter 9 presents formulas to directly compute these two basic parameters (and some PWB vendors offer software that will predict them). Also, as shown subsequently, the reciprocity principle may be used to find the inductance if the capacitance is known.

Reference [18] also gives an equation to compute the mutual inductance (discussed subsequently in Section 4.5) between two microstrips [reproduced in (4.17)].

$$L_{mms} = \frac{\mu_0 \mu_r}{4\pi} \ln\left(1 + \left(\frac{2h}{d}\right)^2\right) \qquad (4.17)$$

The topology for (4.16) and (4.17) is shown in Figure 4.11.

Equation (4.17) properly reports coupling to within 10% if the center-to-center separation (d) is large (three times the trace width or greater), but the error grows to as much as 15% for closer separations.

Equations (4.16) and (4.17) are useful in estimating the inductance of short stubs connecting surface mount pads.

4.6.5 Inductance of Stripline

Equation (4.18) is simplified and recast from [18]. It gives the partial self inductance of single or multiple adjacent striplines. Here the fringing factor curve given in [18] has been numerically fitted for (4.18). Due to the proximity effect, this equation is most accurate for single lines or when adjacent lines are widely separated.

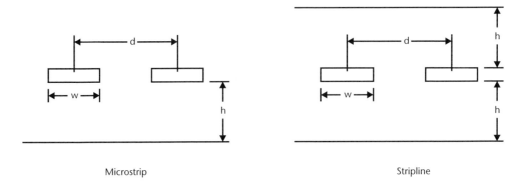

Microstrip   Stripline

Figure 4.11   Topology for (4.16) through (4.19).

$$L_{ssl} = K_{sl} \frac{h}{w + 1.043h} \qquad (4.18)$$

where $K_{sl} = 2\pi \times 10^{-7}$ in meters, or about 15.96 nH/in.

The topology for (4.18) was shown in Figure 4.11.

Typically (4.18) underestimates the inductance of a single conductor by less than 1% for traces with $\frac{w}{h}$ ratios close to one. This grows to 3% for ratios under two and exceeds 7.5% for larger ratios. Assuming FR4 and a 5-mil-wide, half-ounce copper trace, this corresponds to accuracies better than 1% for impedances of 50Ω and below, approximately 3% for impedances between 60Ω and 70Ω, and 7.5% or higher for impedances greater than 70Ω. As is the case for microstrip, as an alternative the inductance may be found once the capacitance is known or the impedance and time of flight have been determined.

Equations are available to compute mutual inductance between striplines, but the accuracy is poor (generally in excess of 20% for widely spaced traces and exceeding 50% when traces are close by). With that caution in mind, (4.19) gives the mutual inductance between two striplines [18].

$$L_{msl} = \frac{\mu_0 \mu_r}{4\pi} \left(\frac{h}{d}\right)^2 \qquad (4.19)$$

4.7   Summary

Inductance causes a voltage drop proportional to the rate of change of the current flowing through the inductance [see (4.4)] and gives rise to an inductive reactance [described by (4.6)].

The path taken by a signal to reach the load and return to the source defines the signal's loop area. The area of this loop determines the inductance of the loop, with larger loops having higher inductance. The concept of partial inductance, described in Section 4.4, allows for the computation of inductance without prior knowledge of the total path taken by the current. Two-dimensional field-solving software creates partial inductance models.

Great attention to the location and consistency of return paths is necessary when connecting several of these models together in a circuit simulator. Reducing the height above a ground plane (the return path) reduces the loop inductance of a wire or trace and lowers its inductance.

The external inductance of a round wire decreases as the natural log of increasing radius and decreasing height above its return [from (4.13)]. Inductance decreases linearly with increasing width and decreasing height for a rectangular conductor, as shown by (4.15).

Mutual inductance relates the magnetic coupling between circuits and can be expressed as an inductance or [from (4.10)] may be described in terms of a coupling coefficient k (as preferred by many circuit simulators). Mutual inductance contributes to crosstalk, but, used properly, it's capable of lowering interconnect loop inductance (e.g., when connecting decoupling capacitors to power and ground planes). However, from (4.12), mutual inductance can cause higher than expected inductance when placing multiple connections in parallel (such as connector pins or vias).

The reciprocity principal fundamentally relates capacitance and inductance in transmission lines propagating energy by TEM (the transmission line propagation mode). This allows for the computation of either capacitance or inductance if the other is known [see (4.3)] and leads directly into the computation of a transmission line's loss factor, time of flight, and impedance (topics covered in Chapter 5).

An inductance matrix shows the relationship between the self inductances (represented on the main diagonal) and the mutual (coupling) inductances (the off-diagonal terms).

References

[1] Miner, G. F., *Lines and Electromagnetic Fields for Engineers*, New York: Oxford University Press, 1996, p. 461.

[2] Ramo, S., J. Whinnery, and T. Van Duzer, *Fields and Waves in Communication Electronics*, 3rd Ed., New York: John Wiley and Sons, 1994.

[3] Skilling, H. H., *Fundamentals of Electric Waves*: New York, John Wiley and Sons, 1967.

[4] Hammond, P., *Electromagnetism for Engineers*, 3rd Ed., Oxford: Pergamon Press, 1986, pp. 12–13.

[5] Sibley, M., *Introduction to Electromagnetism*, London: Arnold Press, 1996, pp. 64–119.

[6] Matick, R. E., *Transmission Lines for Digital and Communication Networks*, New York: McGraw Hill Company, 1969.

[7] Ruehll, A. E., "Inductance Calculations in a Complex Integrated Circuit Environment," *IBM J. Res. Development*, Vol. 16, September 1972, pp. 470–481.

[8] Grover, F. W., *Inductance Calculations: Working Formulas and Tables*, New York: Dover Publications, 1962.

[9] Young, B., *Digital Signal Integrity*, Englewood Cliffs, NJ: PTR Prentice Hall, 2001.

[10] Paul, C. R., *Analysis of Multiconductor Transmission Lines*, New York: John Wiley and Sons, 1994.

[11] Skilling, H. H., *Electrical Engineering Circuits, 2nd Edition*, New York: John Wiley & Sons, 1965.

[12] Apostol, T., *Linear Algebra: A First Course with Applications to Differential Equations*, New York: Wiley-Interscience, 1997.

[13]   Mathsoft Engineering & Education, Inc., Cambridge, MA.

[14]   The Math Works, Inc., Natick, MA.

[15]   Scott, R. E., *Linear Circuits*: Reading, MA: Addison Wesley, 1960.

[16]   Walker, C. S., *Capacitance, Inductance, and Crosstalk Analysis*, Boston: Artech House, 1990, p. 88.

[17]   Sibley, M., *Introduction to Electromagnetism*, London: Arnold Press, 1996, p. 102.

[18]   Walker, C. S., *Capacitance, Inductance, and Crosstalk Analysis*, Boston: Artech House, 1990, pp. 110–115.

CHAPTER 5

Transmission Lines

5.1 Introduction

Lossy transmission lines are the norm on circuit boards, especially when signaling over narrow trace at high frequencies, where skin effect and dielectric losses cause signal distortion. As we'll see in this chapter, at high frequencies the distortion is chiefly caused by unequal attenuation of the signal's harmonics, but phase distortion is the principal cause at lower frequencies. The attenuation is caused by losses due to the series resistance in the conductor and by shunt losses due to the dielectric. The calculations presented in the early sections lump these losses together and involve the use of complex numbers, but simplifications that avoid the use of imaginary numbers and separate out the resistive losses from the dielectric losses are later shown.

Although rectangular waveforms are usually of most interest to the digital circuit designer, the bulk of this chapter focuses on the treatment lossy lines give to sinusoids at single frequencies. This is appropriate because rectangular waves are made up of many single frequency harmonics, and the way each of those harmonics is treated as a pulse travels down a lossy line determines its final wave shape once it arrives at the load. The harmonics reassembly and the effect distorted pulses have on signaling is presented in Chapter 7.

This chapter begins by using ideas from circuit and network theory to analyze a lossy transmission line circuit model. This prepares the way for the discussion in Section 5.4 on traveling waves. The study of traveling waves can become mired in mathematics, but most of that has been sidestepped in this chapter. Instead, as is usual throughout this book, the aim has been to provide enough mathematics to allow an engineer to make hand calculations or to explain results from field-solving or circuit-simulator software. Those wishing a detailed mathematical treatment are referred to the references.

5.2 General Circuit Model of a Lossy Transmission Line

As described in Chapter 4, a TEM transmission line consists of one or more signal lines and a return. When a signal propagates down a transmission line, a time-dependent voltage difference exists between the signal wire and its return, and equal but opposite currents flow along them. The two conductors guide the electric and magnetic fields.

A general circuit model of a TEM transmission line appears in Figure 5.1. As shown, it's made up of many small resistor, inductor, conductance and capacitance (RLGC) segments chained together to represent the entire length of line. Figure 5.1(a) shows the signal and return lines, with each having a resistance (R) and inductance (L).

The signal and return are separated by a dielectric, so a capacitance (C) appears between them. Because the dielectric is not perfect, a shunt loss element (G) appears across the capacitor. The inductance models the energy contained in the magnetic field, while the capacitance models the electric field energy. The series resistance represents the series losses, and the conductance represents the dielectric losses. These elements are smoothly distributed along the length of an actual transmission line, but in the model they appear in lumps representing a small section (Δx) of line. The sections (*lumps*) must be very small (here defined as only a fraction of a wavelength) to give the appearance of a continuous, smooth line.

The resistance and inductance of the return wire may be "folded into" the signal wire, as shown in Figure 5.1(b). This topology is sometimes called a RLGC model.

The distributed RLGC model may be viewed as a chain of an infinite number of π or T sections, with each section representing a very small segment of transmission line. In Figure 5.2, T sections are used with the series R, with L elements equally divided in each arm.

The circuit in Figure 5.2 is a lowpass ladder filter made up of *constant-k* sections, and it's appropriate to use filter theory to determine the circuit's impedance, delay, and attenuation characteristics. To do so, it's first necessary to group the series and shunt impedances, as shown in Figure 5.3.

In transmission line work it's customary to represent the series element (Z_1) as an impedance (Z, with ohms as units, not to be confused with Z_o, the *characteristic impedance* described next) and the shunt (Z_2) as an admittance (Y, the inverse of impedance, with units of siemens). Frequency is expressed in radians/sec (ω) rather than hertz. With this in mind:

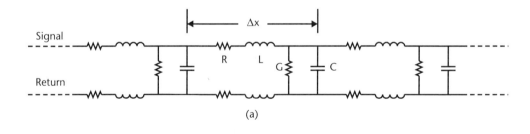

(a)

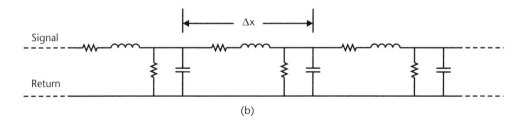

(b)

Figure 5.1 RLGC transmission line models: (a) general model, and (b) return R,L folded into signal.

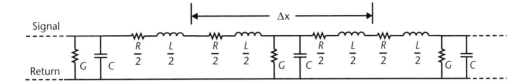

Figure 5.2 T network representation of a transmission line.

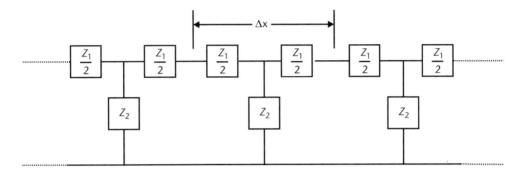

Figure 5.3 Transmission line generalized model.

$$\omega = 2\pi f \tag{5.1}$$

$$Z_1 \equiv R + j\omega L \tag{5.2}$$

$$Y \equiv \frac{1}{Z_2} = G + j\omega C \tag{5.3}$$

where j is the imaginary operator $\left(\text{equal to } \sqrt{-1}\right)$.

From network theory, Figure 5.3 has three properties of interest:

- Its *characteristic impedance* (Z_o) is the impedance necessary to properly match the line so that reflections are not produced when a wave reaches the line's end. When the line is properly terminated, the impedance also determines the relationship between the voltage and current waves traveling down the line.
- It is a lowpass filter: Higher frequencies will be attenuated more than lower frequency ones. The *attenuation constant* (α, units of nepers or decibels) describes the amount of attenuation at each frequency.
- It introduces a phase shift between the waveform launched at the input and the waveform recovered at the output. The phase shift is defined by the *phase constant* (β, with units of radians). As we'll see, this phase shift represents a time delay (td).

The attenuation and phase constants are specified on a per–unit length basis and are often jointly represented by a single value called the *propagation constant*

(γ). The delay is specified per unit length, while the impedance has a value that is independent of length.

5.2.1 Relationship Between ωL and R

The imaginary terms in (5.2) and (5.3) suggest the impedance and propagation constant will be complex numbers, made up of real and imaginary parts, and in general this is true.

However, as we'll see, for PWB trace at very high frequencies, the imaginary terms dominate the real terms, allowing them to be ignored. The $j\omega$ terms then cancel, so that with some approximations it's possible to develop equations for Zo, α, and β that do not use them. In order to do so, it's necessary to determine the frequencies where ωL is larger than R and ωC is larger than G. We begin with the relationship between ωL and resistance across frequency and then examine ωC and G.

Chapters 2 and 4 showed that at high frequency, the series resistance increases as the square root of frequency because of the skin effect, but the inductance remains constant.

At low frequencies ω is small, making ωL small, so at some arbitrarily low frequency ωL is less than the conductor's dc resistance. However, as the frequency increases, ω increases linearly, while R is only increasing as the square root. This means ωL will gradually overtake R, until finally the frequency becomes high enough for ωL to exceed (or even greatly exceed) R.

At frequencies where the skin effect is well developed [F_{skin}, see (2.12) and (2.13)] appearing in Chapter 2), ωL is much greater than R, even for narrow, low-impedance microstrip. This is the worst case for comparing R to ωL across frequency because in general, narrow, low-impedance trace has higher loop ac resistance and lower inductance than wide, higher impedance trace. Such a low-inductance, high-resistance trace will require a higher frequency before ωL is larger than R.

The ratio of ωL to R across frequency is shown in Figure 5.4 for four stripline and microstrip traces. This is example data for half-ounce copper trace on FR4 with a copper return plane. The traces are either 4 mils or 10 mils wide. It's apparent that ωL exceeds R for frequencies in the 10s of MHZ region, even for narrow PWB trace, but it's not until the frequency is in the 100- to 200-MHz range before $\omega L >> R$ for all of the traces shown. The narrow, 50-Ω microstrip (ms2) is the most resistive and thus requires the highest frequency before R exceeds ωL.

A laminate having a lower ε_r than that used to create Figure 5.4 will shift the curves down, requiring a higher frequency before ωL exceeds R.

It's evident in Figure 5.4 that microstrip requires a frequency higher than stripline before ωL exceeds R. This is because for a given impedance, width, and thickness, microstrip trace has both a lower loop inductance and higher loop resistance than stripline (one return path versus two in parallel, hence higher return path resistance).

5.2.2 Relationship Between ωC and G

We now turn our attention to the relationship between ωC and G, with the goal of showing that at high frequencies, ωC is much greater than G. Notice that in this discussion ωC is just the product of the frequency and capacitance and is not the capacitive reactance X_c discussed in (3.10) in Chapter 3.

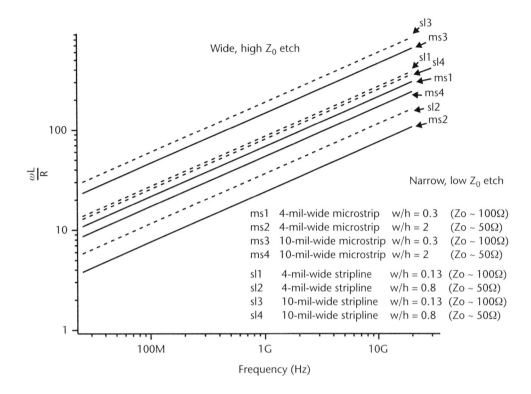

Figure 5.4 R and ωL for narrow and wide microstrip and stripline trace on FR4.

Recall from (3.15) in Chapter 3, that *tan* (δ) (the loss tangent) relates the dielectric losses G to capacitance as $G = \omega C \tan(\delta)$. Because the dielectric losses in PWB laminates are much less than one, and because they do not increase rapidly with frequency (from Table 3.2, FR4 has a *tan* [δ] value of 0.015 at 1 MHz and 0.025 at 1 GHz, for example), the quantity ωC will always be greater than G except at very low frequencies where ω is very small. In fact, for FR4 at 1 MHz and above, ωC will be at least 40 times larger than G at high frequencies and is well over 100 times greater for higher performance dielectrics.

Sections 5.2.1 and 5.2.2 have shown that when signaling over PWB trace at high frequencies, the inductive and capacitive reactances dominate the resistive and dielectric losses. This simplifies the mathematics and will lead to straightforward equations for impedance, loss, and phase shift. But, even though ωC and ωL dominate the losses represented by the R and G terms, the losses can't be ignored. In fact, they are the cause of signal distortion, as we'll see next.

5.3 Impedance

Applying a voltage to the transmission line shown in Figure 5.2 causes current to flow as capacitor C charges. A much smaller leakage current is also drawn by G. Assuming no reflections are present on the line, the voltage-to-current ratio is called the line's *characteristic impedance* (Z_0, with units of ohms, Ω). The characteristic impedance is independent of the line's length and has the same value everywhere along a uniform line, regardless of its length.

5.3.1 Calculating Impedance

Using network analysis, it can be shown (see [1], for example) that if the lumps in Figure 5.3 are small, the characteristic impedance is related to Z_1 and Z_2 as shown in (5.4):

$$Z_0 = \sqrt{Z_1 Z_2} \tag{5.4}$$

Combining (5.2) and (5.3) into (5.4) yields Z_0 in terms of the transmission line's distributed R, L, G, and C components (5.5):

$$Z_0 = \sqrt{Z_1 Z_2} = \sqrt{\frac{Z_1}{Y}} = \sqrt{\frac{R + j\omega L}{G + j\omega C}} \tag{5.5}$$

In those cases where the frequency is high enough so that ωL is larger than ωR and C is larger than G, (5.5) reduces to the familiar equation for the impedance of a lossless transmission line (5.6):

$$Z_0 = \sqrt{\frac{R + j\omega L}{G + j\omega C}} \approx \sqrt{\frac{j\omega L}{j\omega C}} = \sqrt{\frac{L}{C}} \tag{5.6}$$

Example 5.1 compares results from (5.5) and the approximation in (5.6).

Example 5.1

A 5-mil-wide stripline built on FR4 has the following parameters per inch at 100 MHz:

$R = 422$ mΩ

$G = 38$ μS

$L = 10.8$ nH

$C = 3$ pF

Compute the impedance for this transmission line using (5.5) and (5.6).

Solution

From (5.1) $\omega = 6.28 \times 10^8$ rad/sec at 100 MHz.
    Using (5.5):

$$Z_0 = \sqrt{\frac{R + j\omega L}{G + j\omega C}} = \sqrt{\frac{422 \ m\Omega + j\omega 10.8 \ \text{nH}}{38 \ \mu S + j\omega 3 \ \text{pF}}} = 60 - 12.6j \ \Omega$$

At 100 MHz, this transmission line is seen to have both real and imaginary parts to its impedance. The 60-Ω real portion is the resistive part and the imaginary portion (-1.26 $j\Omega$) represents a small capacitive reactance (inductive reactance would be shown as positive) and so indicates that the voltage and current are propagating out of phase. In this case, converting from the ($60-1.26j$) rectangular form to polar form (by taking the arctangent of *imag/real*) yields an angle of $-1.2°$. That is, the voltage lags the current by $1.2°$.

Such a small angle shows the nearly perfect alignment between the voltage and current waves, just as they would be if the line's impedance were purely resistive. Because of this, the lossy impedance calculation given by (5.5) should closely match the lossless impedance given by (5.6). This is in fact the case:

From (5.6)
$$Z_0 = \sqrt{\frac{L}{C}} = \sqrt{\frac{10.8 \text{ nH}}{3 \text{ pF}}} = 60\Omega$$

In this example, the complex portion of the impedance is small enough to be ignored, and the line has an impedance that is nearly purely resistive. This will generally be the case for PWB microstrip and stripline and is especially so as frequency increases. But this assumption does not always hold for thin, narrow traces, especially at low frequency where the resistance is high and ωL is small. This type of trace can be found in some micro packages (where the traces are essentially thin films and have high resistance) or some flexible tape type interconnects that may have high inductance. Depending on frequency, the impedance of these types of trace can have a significant imaginary component, making (5.5) more appropriate than (5.6).

5.4 Traveling Waves

Generally in digital systems, the signals start out essentially rectangular in shape but often arrive at the load with rounded corners and reduced in amplitude. Rectangular waveforms are made from the sum of many sine waves (*harmonics*), each having a specific amplitude and frequency relationship with the waveform's fundamental frequency. The frequency content of pulses is discussed in Chapter 7, but for now it's enough to note that for the signal to retain its original shape, the transmission line must attenuate and phase shift each harmonic by the appropriate amount. Otherwise, the original relationship between the harmonics will not be preserved, and the received signal will be a distorted version of the original.

In fact, lossy lines do not treat the harmonics equally, so pulses do arrive at the load distorted. The amount of distortion depends on the line length, as that determines the degree to which the harmonics are exposed to the incorrect phase and amplitude adjustments. This is evident in Figure 5.5, which shows a 6-ns pulse as it appears at various points along a 36-in-long transmission line. It's plain that distortion increases as the pulse makes its way down the line, with the pulse showing progressively more rounding as it travels. This is caused by the upper frequency harmonics being attenuated more severely than the lower frequency ones. Also evident is the loss of height (amplitude) as the pulse travels down the line, and the pulse shows spreading at its base (smearing) as it travels. As discussed in Chapter 9, successive pulses traveling along this line are more likely to interfere with one another as the boundaries between them (bit times) becomes blurred.

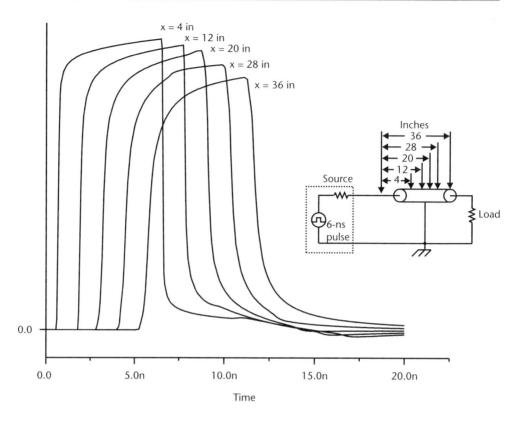

Figure 5.5 A 6-ns pulse propagating down a 36-in transmission line.

To understand the cause of these effects, it's necessary to determine how the transmission line treats each harmonic. For this reason, the remainder of this chapter will focus on the transmission line's response to single frequencies (sinusoids). In Chapter 7, this insight is applied to signaling with pulses.

5.4.1 Propagation Constant

Solving the differential equations relating the voltage across a small segment of transmission line to the current flowing through it yields the equation of a wave traveling along the line's length [2]. For a line with no reflections, the voltage at a distance x is attenuated exponentially from the sending voltage V_s as shown by (5.7):

$$V_x = V_s e^{-\sqrt{YZ}x} \tag{5.7}$$

where Z is the impedance given in (5.2) and Y the admittance given in (5.3).

The quantity $\sqrt{YZ}$ is called the *propagation constant* (this was briefly mentioned at the start of Section 5.2) because it governs the way voltage and current waves propagate down the line. The propagation constant (a misnomer because it varies with frequency) is represented by γ (5.8):

$$\gamma = ZY = \sqrt{(R + j\omega L)(G + j\omega C)} \tag{5.8}$$

The propagation constant is a complex number having two parts: the real portion is the *attenuation constant* (α, with units of neper per unit length), while the imaginary portion is called the *phase constant* (β, units of radians per unit length).

The attenuation constant α determines the way a signal is reduced in amplitude as it propagates down the line, while the phase constant β shows the difference in phase between the voltage at the sending end of the line and at a distance x.

Because it includes ω, (5.8) determines γ at one specific frequency. A pulse contains harmonics of many frequencies. To determine the effect a lossy transmission line has on a pulse, (5.8) must be applied individually to each harmonic. As discussed in Chapter 7, the harmonics are then recombined at the load with unique values of α and β for each frequency to yield a composite waveform at the load [3].

Equation (5.8) produces a total value for α that is the sum of the series resistance and dielectric losses, and it is valid for any TEM transmission line, regardless of the values of R and G. The resistive and dielectric loss contributions are broken out in Section 5.4.4.

Example 5.2

Find γ, α, and β for the transmission line described in Example 5.1

Solution

As the frequency in Example 5.1 is given as 100 MHZ, from (5.1) $\omega = 6.28 \times 10^8$ rad/sec.

A scientific calculator or a scientific software calculation package such as Mathcad [4] or MatLab [5] make it straightforward to compute γ with (5.8):

$$\gamma = \sqrt{(R + j\omega L)(G + j\omega C)} = \sqrt{(11m + j\omega 10.8n)(38\mu + j\omega 3p)} = 0.0047 + 0.113j$$

The real part of γ is α, and the imaginary portion is β. As the RLCG values were all given per inch length of transmission line, the computed values for γ (and so α and β) are the values for 1-in worth of line.

So:

$\alpha = 0.0047$ nep/in
$\beta = 0.113$ radians/in

5.4.2 Phase Shift, Delay, and Wavelength

The phase constant βx shows the phase shift of the voltage (or current) at a point located at a distance x along a transmission line with respect to the sending voltage (or current). A phase shift of 360° (or 2π radians) equals one wavelength and, as shown in Figure 5.6, marks the distance between successive points on the waveform (such as zero crossings).

The wavelength is the distance x required to make the phase angle βx increase by 2π radians. A wavelength is therefore:

$$\lambda = \frac{2\pi}{\beta} \tag{5.9}$$

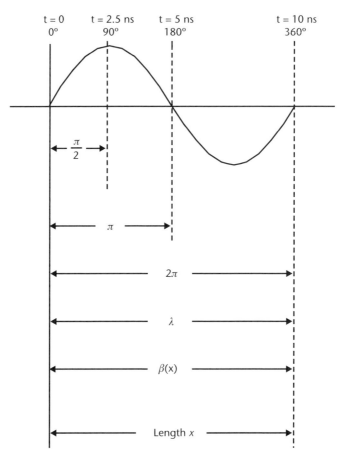

Figure 5.6 Relationship between degrees, radians, phase shift (β), and wavelength (λ).

It's apparent from Figure 5.7 that a phase shift may also be seen as a delay. In fact, expressing β as a sinusoid and taking the derivative with respect to time yields the velocity at which the wave travels down the line [6]:

$$v_P = \frac{\omega}{\beta}$$

(5.10)

From Maxwell's equations, waves propagate along TEM transmission lines with a velocity equal to the speed of light in the dielectric. This leads to particularly useful equations for the velocity of propagation (5.11) and the *guide wavelength* (λ_g) (5.12) [7] for PWB trace where it's assumed the metals are nonmagnetic and so $\mu_r = 1$:

$$v_p = \frac{c}{\sqrt{\varepsilon_r}}$$

(5.11)

$$\lambda_g = \frac{c}{f\sqrt{\varepsilon_{r\_eff}}}$$

(5.12)

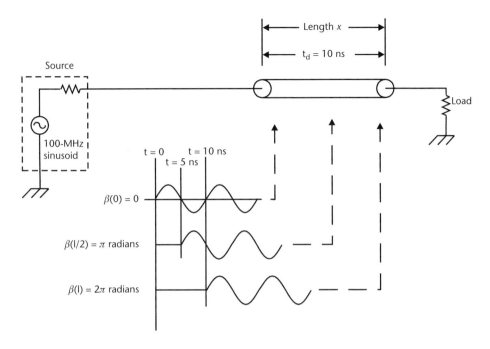

Figure 5.7 Phase shift as a delay.

In (5.11) and (5.12), f is the frequency in hertz, c is the speed of light (3×10^8 m/s or 11.8×10^9 in/sec), and ε_{r_eff} the *effective permittivity*. For stripline, ε_{r_eff} is just the dielectric constant ε_r of the laminate as described in Chapter 3, but this is not the case with microstrip. As discussed in Chapter 9, some of the microstrip electric and magnetic field lines propagate in air as well as the laminate, making ε_{r_eff} lower than ε_r of the laminate itself. In fact, the geometry of a given microstrip determines the value of ε_{r_eff}.

As delay is proportional to the inverse of velocity, the amount of delay a transmission line introduces per distance x is:

$$t_d = \frac{x}{v_p} = \frac{\beta x}{\omega} \tag{5.13}$$

It follows from (5.11) and (5.13) that for stripline, the velocity of propagation (and thus the delay per inch) is the same for all traces, but for microstrip the velocity (and thus the delay) depends on the trace's width and height above a return plane because that's what determines ε_{r_eff}. This is a fundamental difference between stripline and microstrip and is discussed in Chapter 9.

Example 5.3

Using (5.13), what is the time delay of the transmission line in Example 5.2, and what is the delay assuming the trace is a stripline on FR4 with $\varepsilon_r = 4.5$?

Solution

(a) The phase shift (β) and the frequency at which that phase shift is measured (ω) are required to calculate the delay from (5.13). From Example 5.2, the specification is for a 1-in-long line ($x = 1$), making $\beta = 0.113$ radians/inch, and $\omega = 6.28 \times 10^8$ radians/sec.

From (5.13) the delay is then: $t_d = \dfrac{\beta x}{\omega} = \dfrac{0.113 \text{ rad/in}}{6.28 \times 10^8 \text{ rad/sec}} = 180$ ps/in

(b) From (5.11) and (5.13) $t_d = \dfrac{\sqrt{\varepsilon_r}}{c} = \dfrac{\sqrt{4.5}}{11.8 \times 10^9 \text{ in/sec}} = 180$ ps/in

For the signal to appear with the same shape at the end of a transmission line, each harmonic must be delayed by the same amount. From (5.13), β must therefore increase linearly with frequency. Otherwise, t_d would be different for each harmonic, and each would arrive at the load at a different time, improperly altering their phase relationship and yielding a distorted waveform.

Example 5.4

Vias are to be placed every tenth of a wavelength along a 50-Ω stripline fabricated on a laminate having $\varepsilon_r = 4.0$. The highest harmonic has a frequency of 6 GHz. What is the required spacing? What time delay does that spacing represent?

Solution

The wavelength is found directly from (5.12) to be nearly an inch as follows:

$$\lambda_g = \frac{c}{f\sqrt{\varepsilon_{r\_eff}}} = \frac{11.8 \times 10^9 \text{ in/sec}}{6 \times 10^9 \text{ Hz}\sqrt{4.0}} = 0.983 \text{ in}$$

So a via must be placed approximately every 100 mils (one tenth of an inch) to satisfy the 10th-wavelength requirement.

From Example 5.3, $t_d = \dfrac{\sqrt{\varepsilon_r}}{c} = \dfrac{\sqrt{4.0}}{11.8 \times 10^9 \text{ in/sec}} = 169.5$ ps/in, making the delay for 100-mil separation $t_d = 169.5$ ps/in $\times$ 0.1 in $= 17$ ps.

5.4.3   Phase Constant at High Frequencies When *R* and *G* Are Small

It was shown in Section 5.2 that for PWB trace at high frequencies, R and G were small compared to ωL and ωC. This allowed for the simplification of the impedance equation (5.5) to the well known lossless impedance equation given in (5.6). Simplifying the phase constant (5.8) to eliminate the imaginary terms is not as straightforward if R and G are small relative to ωL and ωC but not small enough to ignore. This is the usual case when signaling on PWB trace, especially at high and very high frequencies.

After some involved reductions, [8] eliminates the use of imaginary terms in (5.8) and produces (5.14) for the phase constant at high frequencies when R and G are small but nonzero, and $\omega L > R$ and $\omega C > G$:

$$\beta \approx \omega\sqrt{LC}\left(1 - \frac{RG}{4\omega^2 LC} + \frac{G^2}{8\omega^2 C} + \frac{R^2}{8\omega^2 L^2}\right) \tag{5.14}$$

The time delay for such a line is found by combining (5.14) and (5.13) to produce (5.15):

$$t_d \approx \frac{\beta}{\omega} \sqrt{LC} \left(1 - \frac{RG}{4\omega^2 LC} + \frac{G^2}{8\omega^2 C} + \frac{R^2}{8\omega^2 L^2} \right) \qquad (5.15)$$

Equation (5.15) shows that if R and G are large, each harmonic in a signal will be delayed by a different amount and thus will arrive at the load at different times. Recombining these variously phase shifted harmonics would yield a distorted signal that is not merely a smaller version of the signal launched from the generator.

However, the frequency terms are squared in (5.14) and (5.15), so even at moderate frequencies they dwarf the R and G terms of practical PWB trace. In this case, the R and G terms in (5.14) drop out and the phase constant becomes (5.16):

$$\beta \approx \omega \sqrt{LC} \qquad (5.16)$$

The delay for this line can be found by combining (5.13) and (5.16), this time to form the well-known delay of a lossless line (5.17):

$$t_d = \frac{\beta}{\omega} = \frac{\omega \sqrt{LC}}{\omega} = \sqrt{LC} \qquad (5.17)$$

Because from (5.17) the time delay of a lossless line is not frequency dependent, all of the signal's harmonics will be delayed by the same amount and so will recombine in proper phase at the load. Taken by itself, this suggests that signal distortion, especially at high frequencies, should be negligible. Of course, the opposite is true: signals are significantly distorted by PWB trace, especially so by long trace carrying high-frequency signals. As discussed next, this distortion is chiefly caused by unequal attenuation of each harmonic rather than the improper phase shift at high frequency. But at low frequencies or when signaling over very resistive interconnect, the ω^2 term does not swamp out R, and (5.15) shows that each harmonic will be delayed by a different amount. Such a signal is said to experience *phase distortion*.

5.4.4   Attenuation

Intuitively, signals propagating down lossy transmission lines experience attenuation by an amount that is strongly dependent on the line's length. A line twice as long as another attenuates a signal not a factor of two, but rather by a factor of greater than seven, assuming both lines are properly matched. Matching is important because reflections can change the load voltage, making it appear as if a lossy line has lower (or sometimes greater) attenuation than calculated.

In fact, a sine wave is attenuated exponentially as it travels down a lossy line, as shown in (5.18):

$$V_{fe} = V_{ne} e^{\alpha x} \qquad (5.18)$$

where α is the loss factor found from (5.8) and is expressed in *nepers* (Np, in honor of John Napier, the first developer of logarithms [9]) per unit length. The signal travels a distance x from the *near end* to the *far end* of a line. The value given to α is negative for losses and positive for a gain.

Equation (5.18) is easily solved to find the voltage loss in nepers (5.19):

$$\text{Voltage loss in nepers} = \ln\left(\frac{V_{fe}}{V_{ne}}\right) \tag{5.19}$$

This is illustrated in the following example.

Example 5.5

Equation (5.8) is used to find γ for a certain transmission line at a specific frequency. From that calculation, α is found to be 0.0115 Np/in at that frequency. What is the voltage at the far end at that frequency if the line is 10 in long?

Solution

Because from (5.18) the signal swing reduces as $e^{\alpha x}$, the signal will be reduced to $e^{-0.0115 \times 10} = 0.891$ times its original value. A 1-V input swing would therefore appear on the output with an 891-mV swing. Loss is taken as negative in (5.18) to show the signal attenuates.

It would be convenient to have a way to calculate α directly without first having to calculate γ. In fact, an approximation for α appears as part of the simplification process used previously to obtain β. At high frequencies (where $\omega L \gg R$ and $\omega C \gg G$), [2] shows:

$$\alpha \approx \frac{R}{2Z_0} + \frac{GZ_0}{2} \quad \text{Np/length} \tag{5.20}$$

Example 5.6

Use (5.19) to compute for the transmission line in Example 5.1.

Solution

In Example 5.1, R and G are given at 100 MHz, so α can only be computed at that one specific frequency. Because R and G are specified per inch, the value computed for α will have units of nepers per inch.

$$\alpha \approx \frac{R}{2Z_0} + \frac{GZ_0}{2} = \frac{422m}{2(60)} + \frac{38\mu(60)}{2} = 0.0047 \text{ Np/in}$$

This matches the result from Example 5.2 and shows the good agreement at high frequency between (5.8) and (5.20).

5.4.5 Neper and Decibel Conversion

It's common for voltage loss to be specified in decibels (dB, one tenth of a Bel). Just as nepers express the ratio of the far-end voltage to the near-end voltage on a scale based on natural logarithms, decibels express that ratio on a scale based on common logarithms. Accordingly, (5.21) shows how voltage loss is expressed in decibels:

$$\text{Voltage loss in dB} = 20\log\left(\frac{V_{fe}}{V_{ne}}\right) \qquad (5.21)$$

As with losses expressed in nepers, a negative decibel value represents a loss, while a positive one indicates gain.

The conversion between nepers and decibels is $20\log(e) = 20(0.4343) = 8.686$. That is, multiply the value in nepers by 8.686 to convert it to decibels.

It is often necessary to find the voltage ratio if the loss in decibels (or, sometimes, in nepers) is known. Equation (5.22) shows how to perform this conversion:

$$\text{Voltage ratio} = \frac{V_{fe}}{V_{ne}} = 10^{\frac{db}{20}} = e^{np} \qquad (5.22)$$

The following two-part example illustrates the use of these equations.

Example 5.7

A transmission line connects a source to a load as shown in Figure 5.8. The voltage at the input to the transmission line (V_{ne}) is measured as 800 mV peak/peak, while the voltage at the output (V_{fe}) is measured as 650 mV peak/peak.
(a) What is the transmission line's loss in nepers and decibels?

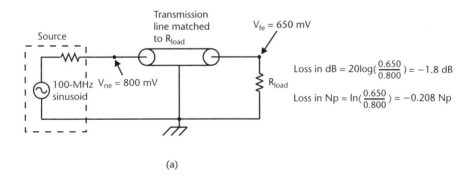

(a)

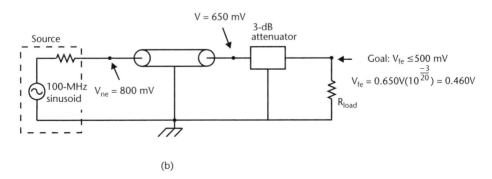

(b)

Figure 5.8   Circuit setup for attenuation calculations in Example 5.7.

Solution

From (5.21), the loss in decibels is $dB = 20 \log\left(\dfrac{0.650}{0.800}\right) = -1.8$ dB, and from (5.19) in nepers it's $nep = \ln\left(\dfrac{0.650}{0.800}\right) = -0.208$ Np.

As a check, -0.208 nep $\times$ 8.686 dB/nep $= -1.8$ dB.
(b) It is desired to reduce the 650-mV signal to below 500 mV by adding a radio frequency (RF) attenuator at the load. Attenuators on 1-dB increments are available from stock. Which one should be selected?

Solution

The required loss is calculated by using (5.21) with $V_{ne} = 650$ mV and $V_{fe} = 500$ mV as -2.28 dB, so a 3-dB attenuator will be selected from stock. From (5.21), a 3-dB loss will reduce the 650-mV signal to $0.65 \times \left(10^{\frac{-3}{20}}\right) = 0.460V$, some 40 mV lower than the minimum requirement.

5.5   Summary and Worked Examples

The following four examples summarize the material in this chapter. Example 5.8 is a simple computation for the loss exhibited by a 12-in-long line when the RLCG parameters are known on a per-inch basis. Example 5.9 shows how to compute RLCG values (and so frequency-dependent loss) at new frequencies when the value is only known at one frequency. In doing so, it draws on material presented in Chapters 2 through 4. Example 5.10 computes propagation delay across frequency for a very resistive line and compares the results to the lossless case. In the final example of the chapter, Example 5.11 breaks out the total loss into its resistive and dielectric loss portions across frequency.

Example 5.8

A 12-in-long transmission line having the characteristics per inch calculated in Example 5.1 is used to connect a source to a 60-Ω load resistor, as indicated in Figure 5.9. When connected as shown, the source output voltage at V_{ne} is 800 mV peak/peak. Determine the voltage at the load (V_{fe}).

Solution

To calculate the far end voltage, it's first necessary to see if the transmission line has the same impedance as its load, because the attenuation and phase equations assume no reflections are present on the line.

In Example 5.1, the line impedance Z_o was found to be very close to 60Ω, so a simple 60-Ω load resistor can be assumed to provide a perfect match. Because there will be no reflections, the attenuation constant can be used to accurately determine the voltage at the load.

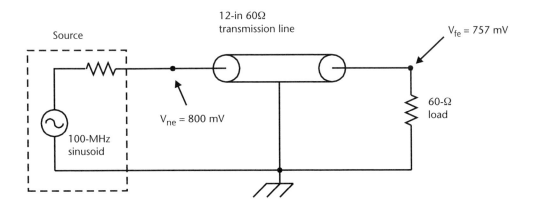

Figure 5.9 Transmission line connecting source to load for Example 5.8.

From Examples 5.2 and 5.6, α = 0.0047 np/in × 8.686 = 0.041 dB/in, so the total loss of the 12-in line will be 12 times that (0.49 dB). From (5.22), the voltage ratio corresponding to a loss of 0.49 dB is 0.945 [noting that the loss is used as a negative value in (5.22)]. If the near-end voltage is 800 mV, the far-end voltage will therefore be reduced to 800 mV × 0.945 = 756 mV.

Alternatively, the same result is obtained by using (5.22) with α in nepers. In this case, α = –0.0047 Np/in, and x is 12.

The second example shows how losses increase with frequency and demonstrates how they can be calculated across frequency even if their characteristics are known at only one frequency.

Example 5.9

Determine how the transmission line in Figure 5.9 attenuates frequencies of 500 MHz, 1 GHz, and 1.5 GHz. Assume the line is 1 in long and is properly matched at all of these frequencies, so no reflections will occur. Further assume the generator output remains "flat" across these frequencies (i.e., stays constant at 800 mV at the frequencies of interest).

Solution

To solve this problem, α must be determined for the three frequencies, but the RLGC values are only known at one frequency. From Example 5.1, the characteristics per inch at 100 MHz are:

R = 422 mΩ

G = 38 μS

L = 10.8 nH

C = 3 pF

We'll first assume the inductance and capacitance values given at 100 MHZ are still valid at the higher frequencies. This is a good physics-based assumption for the inductance and is a good first-order assumption for the capacitance. In Table 3.2, the dielectric constant is seen to decrease as frequency increases, and this will cause C to be lower at the higher frequencies. However, the change is small for a good laminate and typically only results in an impedance change of a few percent when going from 1 MHz to 10 GHz. The change can be greater in lower performance laminates. To simplify things in this example, we'll assume a high-performance laminate and that capacitance remains fixed at 3 pF/in.

Although L and C can be considered constant, the resistance and conductance will vary significantly with frequency. From (3.15), G is seen to increase linearly with frequency, but from Chapter 2 we know that because the frequency is well above F_{skin}, the resistance increases as the square root of frequency.

This means the value of G at 500 MHz will be five times its value at 100 MHz, and R will be $\sqrt{\dfrac{500 \text{ MHz}}{100 \text{ MHz}}} = 2.24$ times as large. Table 5.1 shows the multipliers and corresponding computed values for G and R from 100 MHz to 1.5 GHz. Once G and R are known, (5.1) is used to compute α. That value (and the value when multiplied by 8.686 to convert it into decibels) is also presented in the table. Applying (5.22) yields the value shown for V_{fe} (assuming $V_{ne} = 800$ mV).

The higher frequencies are seen to be attenuated more than the lower ones. If these frequencies represent a signal's harmonics, the attenuation data in Table 5.1 shows an obvious distortion taking place, as the upper harmonics are attenuated far more than the lower frequency ones and thus will disproportionably recombine at the load.

Example 5.10

The transmission line described in Example 5.9 has a constant delay of 180 ps/in for all of the frequencies listed in Table 5.1. Recalculate the delay if R increases tenfold.

Solution

Table 5.2 shows the new results. The time delay (td) is found from (5.8) and (5.13) as was done in Example 5.3.

In this example, the delay is seen to decrease as frequency increases, and the table shows how higher frequencies travel more quickly than lower speed ones on lines having large resistive losses. This is significant to a waveform with a sub nanosecond rise time and will create a jittery, distorted signal. But it's

Table 5.1   Results for Example 5.9

Frequency	F_{mult} (For G)	$\sqrt{F_{mult}}$ (For R)	G (μS) per inch	R(mΩ) per inch	α (Np) per inch	loss (dB) per inch	V_{out} (mV)
100 MHz	1	1	38	422	0.0047	0.040	757
500 MHz	5	2.24	190	945	0.014	0.118	680
1,000 MHz	10	3.16	380	1,333	0.023	0.196	611
1,500 MHz	15	3.87	570	1,633	0.031	0.267	553

Table 5.2   Results for Example 5.10

Frequency	$G(\mu S)$	$R(\Omega)$	td (ps/in)
100 MHz	38	4.22	187.29
500 MHz	190	9.45	181.46
1,000 MHz	380	13.33	180.69
1,500 MHz	570	16.33	180.44

worth noting that this came about because R was quite large compared to ωL. Even a narrow, thin PWB trace is unlikely to have such a large resistance by itself. However, as described in Chapter 2, the switching activity of neighbors sharing a common return path can have the effect of apparently increasing a trace's resistance. Especially for long lines, this effect can be large enough to cause a phase shift in important harmonics, which results in the kind of phase distortion evident in Table 5.2.

Example 5.11

Compare the conductor and dielectric losses for the transmission line of Example 5.9.

Solution

Equation (5.20) breaks out the conductor and dielectric losses. Setting G to zero yields the resistive loss portion, while setting R to zero yields the loss due just to the dielectric. The second column in Table 5.3 shows the total loss results in decibels using the R and G data originally appearing in Table 5.1. The third and fourth columns show the contributions due to resistive loss ($G = 0$) and dielectric loss ($R = 0$).

The total loss is seen to be the sum of the resistive and dielectric losses, and at the lower frequencies resistive loss is higher than the dielectric loss. In this example, that holds until the frequency reaches 1 GHz, at which point the total loss is evenly divided between the two. Dielectric losses are seen to dominate from that frequency on up. For this transmission line, a laminate having a lower loss tangent would improve losses for signal harmonics above 1 GHz.

Table 5.3   Example 5.11 Results

Frequency	Loss (dB) (Total)	Loss (dB) (Resistive Losses, $G = 0$)	Loss (dB) (Dielectric Losses, $R = 0$)
100 MHz	0.040	0.030	0.010
500 MHz	0.118	0.068	0.050
1,000 MHz	0.196	0.097	0.099
1,500 MHz	0.267	0.118	0.149

References

[1]   Terman, Frederick E., *Radio Engineers' Handbook*, New York: McGraw-Hill, 1943, p. 226.

[2]   Sinnema, W., *Electronic Transmission Technology*, Englewood Cliffs, NJ: Prentice Hall, 1988.

[3]   Bertin, C. L., "Transmission-Line Response Using Frequency Techniques," *IBM Journal of Research and Development*, Vol. 8, No. 1, January 1964, pp. 52–63.

[4]   Mathsoft Engineering & Education, Inc., 101 Main Street, Cambridge, MA.

[5]   The Math Works, Inc., Natick, MA.

[6]   Johnson, Walter C., *Transmission Lines and Networks*, McGraw-Hill: New York, 1950.

[7]   Wadell, B. C., *Transmission Line Design Handbook*, Norwood, MA: Artech House, 1991, p. 17.

[8]   Matick, R., *Transmission Line for Digital and Communication Networks*, New York: McGraw-Hill, 1969, pp. 43–49.

[9]   Cajori, F., *A History of Mathematics*, MacMillan: New York, 1919, p. 149.

CHAPTER 6

Return Paths and Power Supply Decoupling

6.1 Introduction

In Chapter 5, a *transmission line* was described as a wire "and its return," with little discussion of just what was meant by a *return*. In fact the return path—the route taken by the current back to the source to complete a series circuit with the load—fundamentally determines how a signal appears at the load. This is because the physical relationship between a signal conductor and its return is what determines the basic electrical characteristics of the line. For example, the line's capacitance and inductance depend directly on the separation between the two conductors forming the transmission line. And in multiple stripline or microstrip situations, the width of the line (and the width of its return) are factors in setting the loop resistance and in determining the magnitude of mutual resistances.

This chapter discusses signals and the return paths used by them under a variety of conditions. Sections 6.2 through 6.4 discuss proper and improper return paths (such as power supply splits and mote crossings). The focus is on single-ended (ground-referenced) signals because these signals are best at illustrating return path concepts.

Signal integrity when routing signal traces through dense pin fields is discussed in Section 6.5, with a discussion of the trade-offs in determining the pad/antipad sizes.

Power supply integrity is a crucial factor in achieving satisfactory signal integrity and has been extensively researched in the literature. This chapter concludes with a review this work in Section 6.6, with emphasis on power supply decoupling, including the use of SPICE-type simulators to model the power supply response in the time and frequency domains.

6.2 Proper Return Paths

Figure 6.1 shows top views of two sets of rectangular traces on an insulating sheet, with no *ground plane* underneath. The signal current is launched down one trace and returns to the source by the second. This is a form of *coplanar* transmission line that is sometimes used in high-speed digital applications. It's used in Figure 6.1 as a convenient way to show the effects when the return path is changed.

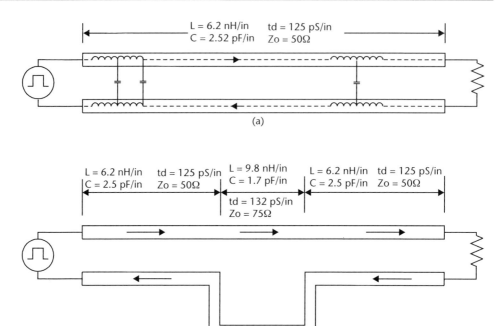

Figure 6.1 Top view showing (a) a signal and its return with uniform spacing, and (b) with an offset in the return path.

In Figure 6.1(a), the signal line and its return are at a constant distance from one another and so the inductance and capacitance per unit length are constant everywhere along the line. Thus, the impedance is uniformly 50Ω.

In Figure 6.1(b) the return trace briefly jogs out, increasing the signal to return separation (and thus inductance) for a portion of the line's length. This increases the loop area between the signal and its return and slightly lowers the capacitance in this region. The net effect is to raise the impedance seen by the signal for this portion of the trace. Of course, this same effect would occur if the signal and return were swapped so that the return is the straight conductor and the signal line has the jog.

This idea is illustrated in Figure 6.2, which shows the side view of a trace changing layers. Here the signal route starts on layer L1, transitions to layer L2, and then

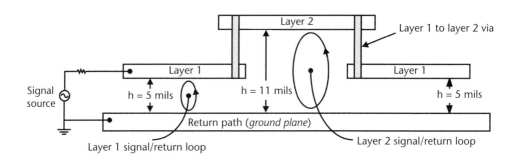

Figure 6.2 Loop area change caused by signal changing layers.

goes back to L1. The loop area between signal and return is clearly larger when the signal is routed on layer L2 than when on L1.

The inductance of the L2 trace will be higher than that of the L1 trace, and the signal will experience an impedance increase (often called an *impedance discontinuity*) for this portion of its length.

6.2.1   Return Paths of Ground-Referenced Signals

It's implied in Figure 6.1 that the current flowing in the return path—the *return current*—is the result of the load connecting the signal to the return path, but this is not so. In fact, coupling between the signal and return causes current to flow in the return before the signal reaches the load. This is seen most easily with capacitive coupling when using microstrip to connect a driver to a load, as shown in a side view in Figure 6.3.

The driver is modeled as a switch S1 connecting 50-Ω pull-up or pull-down resistors to a very long 50-Ω transmission line. The transmission line is shown as a chain of LC PI networks, as described in Chapter 5. For simplicity the series and shunt losses are not shown, and the mutual inductance between the signal line and its return is also not present in the model. The return is shown as being perfect, with no inductance (or resistance). Instead, the signal and return-loop inductances have been combined into (*folded into*) an inductance placed in the signal line only. Two-dimensional field-solving software used to create models used in circuit simulators typically does this. Ammeters M1 and M2 monitor the signal and return-path currents.

Initially S1 connects the output to the return (ground), discharging all of the distributed capacitors. Moving S1 to the position shown at time t = 0 causes a 1.25-V pulse to propagate down the line. Ammeter M1 measures the resulting current to be 25 mA, and even though the voltage wave has yet to reach the lines end, an equal

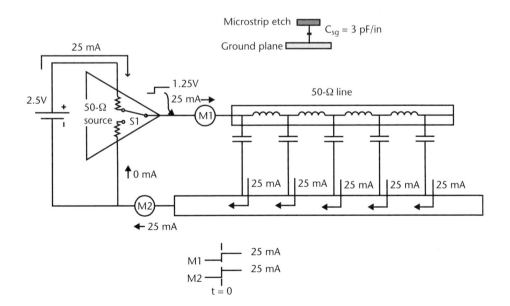

Figure 6.3   Microstrip return path.

current immediately appears on ammeter M2. This comes about because displacement current flows through each capacitor as the voltage wave passes by. This current returns to the source by the ground plane, completing the loop.

Of course, the current launched down the line comes from a power source, shown in Figure 6.3 as a battery having no series resistance or inductance. The 25 mA launched down the line must make its way back to the battery to complete the circuit. The power source and connection are perfect in this example, and the current flows unimpeded to complete the loop as shown.

6.2.2 Stripline

In the microstrip example given in Figure 6.3, the return current flows in the one return path: the single ground plane. The situation is a bit different with stripline because there the trace is sandwiched between two return paths, and so the return current has two routes back to the source. The proportion of current flowing in each depends on the closeness of the trace to each plane. If the trace is centered between the two planes, the return current splits equally between the two because the trace capacitance to each plane is identical. This is not the case with offset stripline, where the trace is closer to one plane than the other. This situation arises in *orthogonal routing* situations, where two stripline layers are formed between two ground planes, with one routing layer carrying signals in an east-west direction and the other layer carrying them in a north-south direction. The return plane closer to the trace will carry proportionally more return current than the more distant plane, as the capacitance between the trace and the close by return plane will be higher. This distinction is unimportant if the return planes are both ground, but it can matter if one plane is a power supply voltage.

6.3 Stripline Routed Between Power and Ground Planes

It's sometimes necessary to create stripline by routing trace between power and ground planes. As explored in this section, the power supply chosen can alter the quality of the signal received at the load. We begin first with the proper case: forming stripline with a power plane that is associated with the signal voltage. The following section discusses how using a plane connected to an unrelated voltage can adversely affect signal quality and examines the role of decoupling capacitance in improving return path quality.

6.3.1 When Power Plane Voltage Is the Same as Signal Voltage

A 50-Ω driver connected to a 2.5-V source and a long 50-Ω transmission line is shown in Figure 6.4. The transmission line is stripline, formed by ground plane on one side and a 2.5-V power plane on the other. The end of the line is simply open circuited: there is no load at the far end. Although critical at high frequencies, decoupling capacitance between the power and ground planes is not included in this model. This topic is explored in Section 6.6.

As with the microstrip of Figure 6.3, the distributed capacitance is shown as many lumped capacitors in parallel. The total capacitance in each lump is C. In this

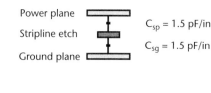

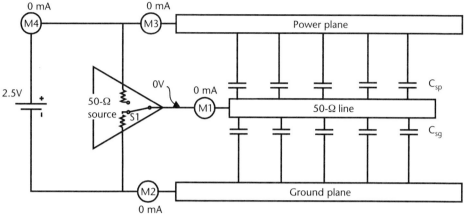

Figure 6.4   Stripline between power and ground planes.

example the trace is centered between the two returns, so half the capacitance goes to each plane and is shown as $\dfrac{C}{2}$. Switch S1 has been in the position shown for long enough to totally discharge capacitors C_{sg} and to totally charge capacitors C_{sp}. Therefore there is no longer any displacement current flowing and the circuit is at rest.

Moving S1 to connect the pull-up resistor as depicted in Figure 6.5 causes M1 to register the expected 25 mA as the 1.25-V wave is launched down the 50-Ω transmission line, but meter M4 indicates only half as much: 12.5 mA. This is explained by the current flow sketched in Figure 6.5.

The total current launched down the line is the sum of the current flowing as C_{sp} discharges plus the current required to charge C_{sg}. Because these capacitances have the same value and because in this example the pull-up and pull-down impedances are equal, the displacement currents flowing in C_{sg} and C_{sp} are the same, and, as shown, half the total current I flows to each plane. Because V1 only charges C_{sg}, it only has to source half the total current: $\dfrac{I}{2}$. Eventually all of the C_{sg} capacitors become fully charged to 2.5V, and all the C_{sp} capacitors fully discharge. The current through meter M1 then falls to zero and the transmission lines voltage becomes the same as V1: 2.5V. From a transmission-line perspective, this occurs when the reflection voltage from the transmission line's open far end has come back to the near end. At that point, the line is at a steady state with 2.5V everywhere along the line.

Connecting S1 to the ground plane as shown in Figure 6.6 discharges the line.

Once again, meter M1 registers the expected current: −25 mA (the negative sign indicating that the driver is now sinking current) as the line is discharged from the 2.5-V steady state. As is usual with transmission lines, voltage divider action between the driver's impedance and the line's impedance initially sets the near-end

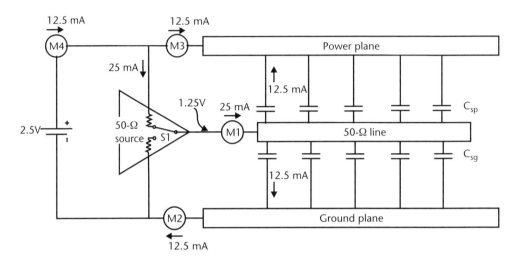

Figure 6.5   Launching a positive voltage step down a 50-Ω stripline.

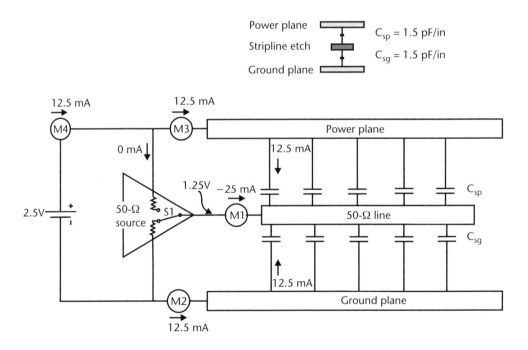

Figure 6.6   Discharging a 50-Ω stripline.

voltage to 1.25V (and not to zero). And, once again, meter M4 shows that V1 is sourcing half that current: 12.5 mA. The currents sketched in Figure 6.6 show why V1 should be sourcing current when the line is being discharged. This case is really the dual of the charging case, with C_{sp} being charged by current sourced from V1 and C_{sg} simply being discharged directly by the driver with no path through V1.

6.3.2   When Power Plane Voltage Differs from Signal Voltage

The situation becomes more complicated when the power plane forming the stripline is not connected directly to the power source of the driver. This case is shown in Figure 6.7, where a 2.5-V signal is routed between ground and a plane carrying 3.3V. This type of situation sometimes arises on PWB cards having multiple power supply voltages and not enough routing layers to keep the signal referenced to the proper power plane. As was the case for the stripline in Section 6.3.1, in this example the end of the line is simply open circuited: there is no load at the far end, and once again decoupling capacitance between the power and ground planes is not included in this model.

The return current flow is the same as in the previous example when discharging the load, but the flow is subtly different when the line is being charged. The currents for this case are sketched in Figure 6.7. Power supplies V1 and V2 are seen to be perfect, having no series resistance or inductance, and they are connected through a perfect ground plane. The return current flows freely through V2 to V1 at all frequencies, and the return loop is completed without difficulties, just as it was in Section 6.3.1. From this it would seem that using an unrelated power plane to form a stripline could be done without adversely affecting the signal quality, but this is not so. In fact, without care this situation leads to poor signal quality and increased jitter because any inductance in V1 or V2, or in the connections between them, contributes to choking of high-frequency components of the return current. Careful attention to the amount and placement of decoupling capacitance is required to make the return path through the power system operate satisfactorily at very high

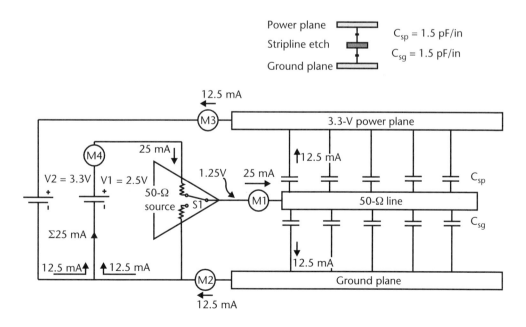

Figure 6.7   Charging a stripline formed by ground and an unrelated power plane.

frequencies. This capacitance comes in two forms: that which occurs naturally between power and ground planes, and that which occurs from the decoupling capacitors placed intentionally on the PWB's surface.

6.3.3   Power System Inductance

In practical systems, the power supply interconnect has impedance that varies with frequency. The frequency characteristic of a power/ground plane pair with and without decoupling capacitors is analyzed in Section 6.6. For the purposes of this section, we note that the power supplies previously assumed perfect actually have substantial inductance associated with them. This represents an impedance that increases with frequency and thus blocks high-frequency return currents from flowing through the supplies. The bulk decoupling present in the power supply (such as a switching or linear regulator locally placed on a PWB) has a low enough impedance to properly steer the low-frequency return currents. This leaves it to the small-valued, low-inductance decoupling capacitors placed on the PWB's surface to steer the high-frequency currents. The efficacy of these capacitors depends of their physical placement and on layout-related parasitic inductance. As discussed in Section 6.6, this inductance creates an resistor, inductor, capacitor (RLC) tank that will resonate. Once past series resonance, the capacitor's impedance increases with frequency, and the capacitor is said to be *inductive*. Fortunately, the power/ground planes themselves are low-inductance structures that, in general, will series resonate at frequencies higher than the decoupling capacitors. In fact, depending on the decoupling layout, it's possible for the highest frequency currents to be returned by this path [1].

A circuit schematic showing the current flow in the presence of decoupling capacitance appears in Figure 6.8. Inductor L_{supply} is large enough to prevent high-frequency return currents from flowing through the power supply. As shown,

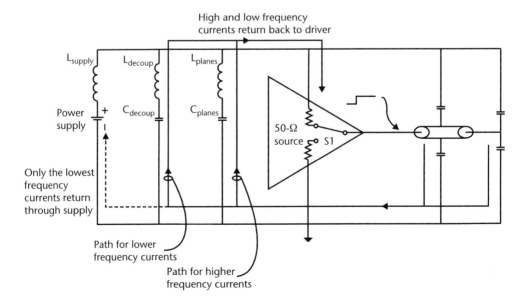

Figure 6.8   Return paths in the presence of power supply inductance.

discrete capacitance added to bypass the supply (C_{dcoup}) steer current around that inductance, but they have parasitic inductance associated with them, too, and so are decreasingly effective as frequency increases. It's left to the capacitance appearing between the power and ground planes (C_{plane}) to provide a return for the highest frequency currents.

6.4   Split Planes, Motes, and Layer Changes

In describing proper return paths, the previous section laid the groundwork for this section: the discussion of PWB layout that forces return currents to flow back to the source by nonideal paths. These incorrect paths are usually referred to by the generic term *split planes*, but they also include mote crossings and layer changes. Return currents through connectors also fall into this category. Discussion of that important topic is deferred to Section 6.5.

6.4.1   Motes

Motes (or *power islands*) are perhaps the most obvious split-plane situation. A *mote* is defined as a total break in the copper plane, forming an isolated region. This technique is often used to form unique power islands that connect either to a voltage different from the rest of the plane or to the same voltage through a PI filter (formed with a ferrite or inductor with shunt capacitors). This second situation arises when it's necessary to provide filtered power to the pins of an integrated circuit, as might happen when deriving an isolated input/output (I/O) power supply from a common supply of the same voltage. Top and side views of a signal crossing a mote formed with a PI filter appear in Figure 6.9. In this example, the signal trace is microstrip running on the surface, with a pair of power and ground planes underneath.

A signal is seen being driven from an I/O buffer powered by the filtered supply formed by an inductor/capacitor PI filter. The side view in Figure 6.9 shows how the mote interrupts the power supply portion of the return path. As shown, the ground plane runs underneath in this example and is assumed to be unbroken.

Within the island region, the signal is microstrip referenced to the isolated 2.5-V power supply formed by the power island. The portion of the signal crossing the mote is referenced to the ground plane and in effect becomes a high-impedance microstrip. However, the mote is likely to be narrow enough so that this portion of the line is electrically small. If so, this section is more like a lumped series inductor and small shunt capacitor than a transmission line [2]. But regardless of the mote's width, the impedance increases in this region because the return path loop area has increased (causing the inductance to increase [3]). Also, the capacitance is lowered ($C_{mg} < C_{ms}$) because in this region the trace capacitively couples to the ground plane, which is further away than the power plane. The final portion of line is referenced to the main power plane. Although not important in this example, the edges of the planes forming the mote and main 2.5-V power plane facing each other create small a coupling capacitance (C_m) that is in parallel with the ferrite L1. As we'll see later in this section, making the mote larger improves isolation by reducing this capacitance.

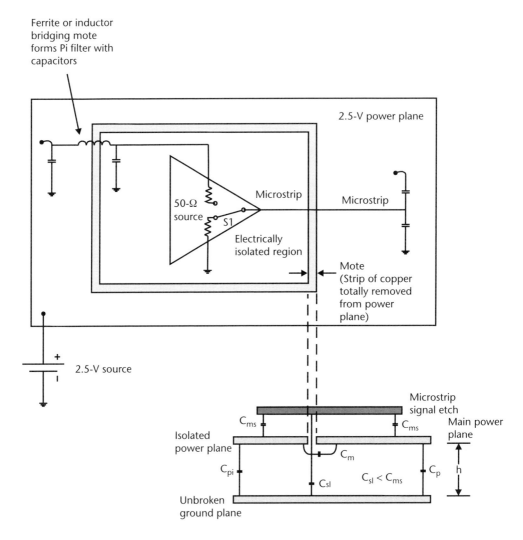

Figure 6.9 Signal crossing a mote.

Figure 6.10 is a time-domain reflectometry (TDR) [4] plot of a 50-Ω (nominal) microstrip conductor crossing a large split in the power plane and entering a mote region as depicted in Figure 6.9. In Figure 6.10, the TDR pulse is launched from the 2.5-V side of the mote and travels down the trace, crosses over the cut in the 2.5-V plane, and enters the mote proper. The TDR was taken on a board having no components: the driver, capacitors, and inductor shown in Figure 6.9 are not present. Initially the TDR pulse is referenced to the 2.5-V power plane and has a 53-Ω impedance. An increase in inductance and reduction in capacitance causes the impedance to increase to nearly 70Ω when the signal crosses over the void separating the 2.5-V plane from the quiet 2.5-V plane within the mote. The impedance eventually returns to about 60Ω once the signal is within the mote region. Significantly, even though the power island is disconnected from the main 2.5-V plane, the impedance doesn't remain at 70Ω once the signal reaches the power island's interior. This is because the ground plane that runs underneath both the main 2.5-V plane

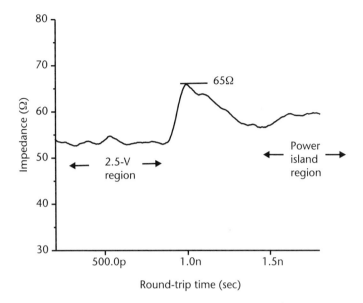

Figure 6.10   TDR of a signal-crossing a mote, as depicted in Figure 6.9, sans driver, L1, and capacitors.

and the quiet 2.5-V plane within the mote capacitively couples the two power regions together (the series connection of C_{pi} and C_p). This indirect connection by way of the ground plane elevates the return path impedance, and so the impedance as looking from the outside to the inside of the mote is higher than the impedance outside the mote.

Rather than focusing on impedance changes, it's also possible to consider the mote crossing as an energy mode conversion process, where the energy initially propagating along the microstrip gets converted into a signal propagating along a slotline-type transmission line [5, 6]. This type of approach is especially useful when both the power and ground planes are split and so the signal runs across a region that is totally void of return path metal.

It's best to move mote-crossing signals to a routing layer that has an unbroken return path, such as stripline formed between two ground planes or a surface-layer microstrip that has a contiguous ground plane underneath. Opto-isolators are very effective in providing total isolation and are a good solution when a limited number of low-speed signals must cross a mote.

Differential signaling is sometimes used to cross motes when higher speed or wide busses make opto-isolators impractical. The idea behind this scheme is that because the voltage at which differential signals switch is relative only to the two leads forming the differential pair (and not with respect to ground), the mote crossing will have less of an impact because each signal in the pair is mistreated identically as the mote is crossed. However, as pointed out by [6], even when signaling differentially, at least some of the return current flows in the ground plane. The magnitude depends on the relative height above the ground plane verses the spacing between the traces forming the differential pair. The result is that the differential impedance greatly changes when the signal crosses over the mote, causing a distorted signal. The impedance change can be somewhat mitigated if the signal is far above the ground plan, but, in general, crossing a mote with a differential pair does not result in satisfactory high-speed signaling.

Up to now, the discussion has centered on what happens to signals when they cross a mote. The interaction of the planes on either side of the mote (the power islands) also requires attention. The edges of the metal planes facing each other across the mote form a capacitor (C_m in Figure 6.9) that provides a coupling path across the mote. However, this aberrant path will only be favored if the interplane capacitance (C_p and C_{pi}) is small relative to C_m. Boards with a thin dielectric between the power and ground planes will have larger C_p than thicker ones, and thus the mote can be made narrow and still achieve adequate isolation. As a general rule, isolation improves with increasing mote width up to about a distance equal to two times the interplane spacing [7] (h in Figure 6.9). Isolation improves slowly for greater separations. However, if the planes resonate, the isolation will be small, especially if both sides of the mote resonate at the same frequency. For this reason, [7] points out that power islands should be different sizes (which is usually the case). Increasing the mote width to greater than $2h$ will help reduce coupled energy.

6.4.2 Layer Changes

If improperly managed, layer changes—the use of vias to connect trace on different layers in a stackup—can alter the return path taken by signal currents and so can adversely affect signal quality.

A signal propagating along a via has currents traveling orthogonal to any planes the via passes through, making the return currents preferentially choose other structures (such as nearby vias) to act as returns. Signals sent along vias can experience significant changes in impedance and increased crosstalk [8] and can be a source of electromagnetic interference (EMI) [9]. For single-ended (*ground-referenced*) signals, placing *companion* vias tied to the reference plane (usually ground) and placed adjacent to the signal via can improve these deleterious effects by providing a local return path.

Differential signals should always change layers by placing the vias side by side so as to maintain the correct differential impedance (such as 100Ω).

6.5 Connectors and Dense Pin Fields

It can be difficult to maintain proper return paths when routing signals through the dense pin field of a fine-pitched connector or ball grid array (BGA)–type micropackage. The signals connecting deep within the pin field must pass by rows of signal and ground pins. This results in coupling that can give rise to undesirable levels of crosstalk, especially if many signals switch simultaneously. This is a special concern for single-ended (ground-referenced) signals, where the crosstalk can give rise to data-dependent jitter on transmitted signals. It's less troublesome with differential signaling. This is because the trace can be routed to pass next to pins forming a differential pair, where one signal in the pair will drive high while the other drives low, effectively canceling (or at least greatly reducing) the coupling. Additionally, if the signals routed through the pin field are themselves differential, the routing can be arraigned so that coupled noise becomes common to both signals forming the differential pair (*common mode* noise). This greatly reduces received jitter.

Signal traces passing by a pin connected to the power or ground planes will have coupling from the trace's edge to the pin. This increases the trace capacitance and lowers the inductance in the segment adjacent to the pin, thereby lowering the trace impedance in that region.

6.5.1   Plane Perforation

As described in Chapter 1, each signal pin that passes through a plane without connecting to it must have a region void of metal (an antipad or clearance region) so that the pin does not short to the plane. As shown in Figure 6.11, these antipads create punchouts (perforations) in the power and ground planes. If the vias are placed close together, these antipads reduce the metal available between vias to carry current. This is sometimes called *ground starvation*. Signals referenced to such a starved area can have increased impedance and coupling (crosstalk) [10].

As was shown in Example 3.2, another effect of these perforations is to reduce the amount of naturally formed decoupling capacitance appearing between a power and ground plane. Although each antipad removes only a small amount of capacitance, as was shown, the aggregate effect over a large board can be surprisingly high.

6.5.2   Antipads

Another concern when routing through dense pin fields is having signals completely or partially cross the antipad clearance region, as shown in Figure 6.12. Electrically this is similar to the mote crossing discussed in Section 6.4.1 in that the signal is no longer referenced to a nearby power or ground plane in this region, so signal impedance will increase.

In Figure 6.12, two 10-mil-wide traces have been routed between connector pins in a dense pin field. A single routing layer is shown with two edge-coupled

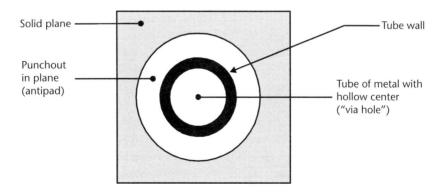

Figure 6.11   Top view of a via passing through a solid plane. Antipads are punchouts in the plane providing clearance between the tube of metal forming the via and the plane the via passes through.

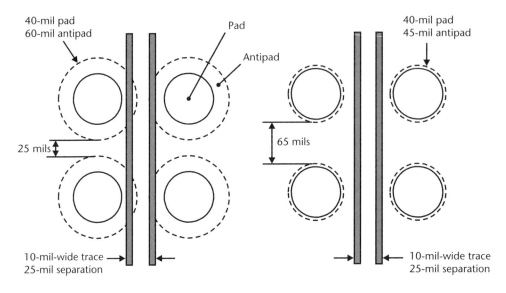

Figure 6.12   Signal routing through a dense pin field. Large antipads cause signal traces to pass partially or fully over the punchout in the plane, increasing inductance and reducing capacitance.

traces, and for simplicity the 40-mil via hole is shown without the walls shown in Figure 6.11. The antipad is actually formed in the planes above and below the routing layer, so for emphasis the antipads are drawn with broken lines. Spacing between signals has been made large in an attempt to maintain a certain differential impedance (such as 100Ω). The traces pass over the antipads when the antipads are large but are well clear of them when the antipads are small.

By routing the signal so that it doesn't cross the antipad, the signal remains properly referenced, and its impedance should not increase. However, the impedance will decrease slightly due to coupling between the trace and connector pins if the trace is routed too close to a row of pins tied to power or ground. Generally the impedance change is small and because the pin diameter is small, the discontinuity is of short duration. Nonetheless it's best to route signals as close to the center of the routing channel as possible, thereby maintaining the greatest distance between the trace and pins.

At first glance, small antipads would appear to be the best from a wiring perspective because, being small, they remove the least amount of copper from the power and ground planes. This is evident in Figure 6.12 and has the benefit of reducing ground starvation effects, as it creates the widest routing channel between pins. However, capacitive coupling from the signal pin in the connector or BGA via to the power or ground plane is highest when the antipad is small. Parasitic capacitance that couples the pin field via to the board's power or ground planes will cause the connection to be more capacitive than expected, lowering its impedance. In fact, a small antipad can cause a connector designed to have a 50-Ω impedance to appear as a much lower impedance.

This is clear in the TDR plots presented in Figure 6.13, which shows the results of a TDR signal launched through the pin field of a 50-Ω connector that is in turn connected to 50-Ω stripline. The stripline does not cross over the antipad; rather, it remains well referenced for its entire length.

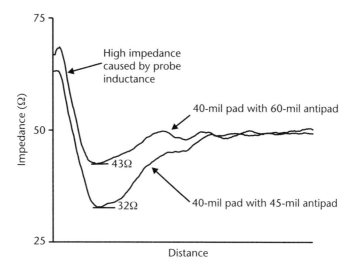

Figure 6.13   TDR of 40-mil via with 45-mil and 60-mil antipad.

Two traces are shown, slightly offset horizontally for clarity. One trace is the measurement of a 40-mil-diameter via with a 45-mil antipad, while the second is a 40-mil via with a 60-mil antipad. The vias are 145 mils in length fabricated on FR4 with ε_r measured as 4.0.

The TDR signal initially enters the connector pin field at the left of the plot, where the impedance spikes up due to probe loading caused by a large loop area formed between the probe and its return. The impedance of the connector pin with the 45-mil antipad is seen to fall to 32Ω and gradually recovers to 50Ω as the signal makes its way down the 50-Ω stripline. This is in contrast with the 43-Ω impedance and quicker recovery of the pin with the 60-mil antipad. The pin having the largest antipad is seen to have the lowest parasitic coupling.

While the smaller antipad can free up routing channels within a pin field and provides more return path metal, it's clear from Figure 6.13 that if made too small it creates a large parasitic capacitance that significantly reduces the connector's pin impedance. On the other hand, making antipads very large can make pin impedance more uniform (especially in thick boards having long vias), but this reduces the number of wiring channels available through the pin field and has the potential for removing so much return path metal that signals are impaired in other ways.

Resolution of this dilemma requires a trade-off unique to each design situation that compares the number of routes required between each pin to the coupling from trace to pin that will occur with that spacing. Another factor is that routing fewer signals between rows of pins will require more routing layers to connect all of the pins, thereby increasing the board's cost and thickness. A thicker board implies a longer via, and longer vias will be subjected to more coupling than shorter ones. This makes it desirable to keep the board as thin as possible and brings us back to where we started because the fewer number of routing layers in the thin boards makes it necessary to route more signals between pins.

The choices are to use narrow trace (increasing resistive losses, as described in Chapters 2 and 5), or to minimize the antipad size and accept the change in impedance, or to add more routing layers to increase the thickness (and cost) of the board.

In the end, the solution typically involves some compromise between all of these factors.

Another powerful design factor that is sometimes overlooked is the choice in laminate. Selecting a laminate having a lower dielectric constant (ε_r) will help reduce the impedance change when routing close to the pins because the coupling capacitance will be reduced. For the same reason, such a laminate also allows the antipads to be smaller before the pin impedance is affected, as coupling between the pin via and ground planes within the board will be less. A beneficial side effect is that for a given impedance, the lower ε_r requires trace to be closer to the ground plane, making the board thinner. This further reduces the via pin field coupling because the via will be shorter. This effect is small for boards with only a few routing layers but is significant in larger stackups. The lower ε_r also allows signals to be routed closer together for a given amount of crosstalk and means differential pairs can be routed closer together and still maintain the proper differential impedance.

This multidimensional problem is best analyzed with a circuit simulator that uses pin and via models created by a three-dimensional field solver. The model should include the amount of metal available in the pin field for return paths (ground starvation) and a good lossy model of the trace connecting to the pins.

Alternatively, a test circuit board can be made that has various pad/antipad and trace width combinations. This approach is very effective if the cost can be justified and if the proper test equipment is available.

6.5.3 Nonfunctional Pads

Sometimes PWB manufacturers include *nonfunctional pads* in via stackups. These are normally sized via pads placed on vias and located on routing layers that do not connect to trace or to other copper such as planes. Their purpose is to better anchor the via into the laminate. They are never placed on power or ground layers because doing so would connect the via to the plane, shorting the signal.

In many situations, these nonfunctional pads are harmless, but they can be troublesome in high-speed signaling, especially in thick multilayer boards that have small antipads. Figure 6.14 shows how the nonfunctional pad capacitively couples to a return plane, lowering the impedance seen by a signal passing along the via.

Initially, the signal is routed on layer L3 but transitions to layer L1 by way of an L1 to L3 via as shown. This is a typical via running through the entire stack of the six-layer board and is not a *blind* or *buried* via that is only long enough to connect together the two layers of interest. A pad is present on each signal layer, even if that layer does not route trace to the via. Pads are seen to be present on layers L1 and L3 as expected, but an unused (nonfunctional) pad is also present on L5, even though trace is not being routed to the via on that layer. As shown, this nonfunctional pad will capacitively couple to planes (shown as ground plans but these could also be power planes) above or below in the stackup. This unwanted capacitance adds capacitive loading to the via, and if the via is long enough (or if the frequency is high enough), the unused portion of the via will appear as a stub.

It's often considered mandatory to remove nonfunctional pads from via stackups on high-speed nets, but this is not always necessary in practice. If the antipads surrounding the vias are large, the coupling between the nonfunctional pad and the

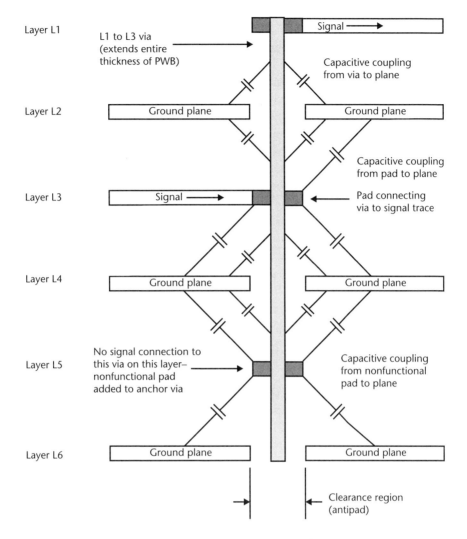

Figure 6.14 Cross-sectional view of a via passing through a six-layer PWB. The coupling experienced by vias (including nonfunctional pads) is clear.

plane will be reduced. Likewise, the coupling capacitance will be reduced if the trace height above the planes is large or if the dielectric constant is small. Some manufacturers will allow designers to specify that nonfunctional pads be forbidden on routing layers that have a power or ground plane on adjacent layers. For example, applying this rule to Figure 6.14 would eliminate the nonfunctional pad on L5 and would eliminate its coupling to the planes on layers L4 and L6.

6.5.4   Guidelines for Routing Through Dense Pin Fields

Simulation or measurement is the best way to determine the optimum trade-off between trace width, separation, and the pad/antipad size for any given high-speed signaling situation. However, the following guidelines are generally applicable and are useful in guiding a simulation strategy.

- Maintain significant numbers of returns through the connector. Provide no less than one return (ground) for every one to two single-ended signals, located in the signal's vicinity. Slower speed signals can afford fewer references, but higher speed signals cannot. Ideally, treat each single-ended signal as if it were a differential signal, with ground being the second half of the diff pair, and route both through the connector. If true differential signaling is employed, provide at least one ground for every diff pair. All high-quality high-performance connector systems provide at least a 1:1 ground-to-signal pair ratio, and some have higher ratios (more grounds than diff pairs). Providing adequate return paths through the connector in this way improves signal integrity, but doing so also provides a low-impedance connection between the ground planes of the two boards being connected. This helps to control EMI [11].

- Maintain a uniform trace impedance through the pin field. If the lines are differential, keep the trace spacing uniform to maintain the proper differential impedance (such as 100Ω). This is often difficult if the pin field is fine pitch or if the trace width has been made wide to reduce resistive losses. In this case, it's permissible to *neck down* the trace width for the short distance the trace is within the pin field (see Figure 6.15). If the necking region length is short enough, the impedance discontinuity will be negligible.

- Route differential pairs side by side (if edge coupled) for as long as possible within the pin field. This helps to keep any noise common mode, improving noise immunity. Always route edge coupled diff pairs on the same routing layer: do not escape on separate layers and then use a via to get both signals on a common layer once outside the pin field. Doing so will compromise noise rejection and increases skew, and the energy reflected from the via will contribute to eye closure. Routing through as dense a pin field as broadside pairs (discussed in Chapter 8) can help alleviate these types of common mode noise problems.

- Route high-speed signals on the outermost connector rows, thereby avoiding routing deep within a pin field. This allows the signals to maintain a uniform spacing for as long as possible and reduces as much as possible the coupling to pins described earlier. However, the signal pins on the outside of the connector may not be as well referenced (shielded) as the innermost ones. If so, routing high-speed signals to the outside pins will increase the likelihood of EMI problems [11].

- Never route trace over an antipad region.

- If the PWB is thick, use large antipads to minimize capacitive coupling between the connector pins and the ground or power planes. Also, consider a low ε_r laminate. Both these things will minimize the impedance discontinuity exhibited by the pin. Note that too large of an antipad can result in ground starvation, degrading signal quality. However, thick boards often have multiple power/ground planes, somewhat mitigating the ground starvation effects. The use of one-ounce copper in the power/ground planes (rather than half-ounce copper often used for signal traces) can also help mitigate this effect.

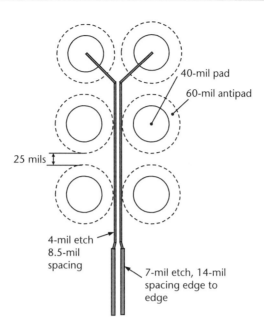

40-mil pad

60-mil antipad

25 mils

4-mil etch
8.5-mil
spacing

7-mil etch, 14-mil
spacing edge to
edge

Figure 6.15   Reducing trace width (necking down) to better route through a connector pin field. Because the wide trace is designed for 50Ω, the impedance of the narrow trace will be higher.

6.6   Power Supply Bypass/Decoupling Capacitance

For the discussion up to now, the power supplies have been assumed to have an impedance low enough at the frequencies of interest to permit return currents to flow as desired. In fact, the connection from an integrated circuit to the power system has substantial inductance, which blocks all but the lowest frequency return currents from flowing through the power sources, as described in Sections 6.2 and 6.3.

Power supply decoupling provides three benefits to a design: (1) improves power supply integrity, (2) improves signal integrity, and (3) helps control EMI. Generally, designs with good power supply and signal integrity have the fewest EMI problems, and designs with good power supply integrity provide the best electrical environment for signal transmission. Proper decoupling—the placement of capacitance between power and the return (usually ground)—can make the difference between a design working marginally and one that is reliable. It's usually difficult to retrofit decoupling capacitance into a marginal design to turn it into one that works reliably. It's always better to properly design in the power supply and return path decoupling.

The EMI benefits are discussed in [12] and are not further discussed here other than to observe that reducing the loop area traversed by signals is beneficial in reducing emissions. As described in Chapter 4, this also improves signal integrity.

Power supply integrity has been extensively studied in the literature [13–18]. A discussion of this work follows, with a focus on multilayer boards incorporating power and ground planes. Double-sided boards that use wide traces to route power to all of the elements are generally not used in high-speed design and so are not discussed here.

6.6.1 Power Supply Integrity

Capacitance used to *decouple* a power supply can be analyzed in either the time domain (the traditional way still often preferred by many engineers) or in the frequency domain. Time-domain analysis is convenient, as results can be readily correlated with oscilloscope measurements, but frequency domain analysis gives insight into the behavior across a spectrum of frequencies. This is helpful when analyzing RF emissions (EMI) and in selecting the type and location of decoupling capacitance.

Viewed in the time domain, decoupling capacitance is a source of charge that can be placed where desired. In the frequency domain, decoupling capacitance lowers the power supply distribution impedance. Thinking about the impedance of the power distribution network makes intuitive sense, as a low-impedance network will have a lower voltage drop than a high-impedance network. This is another way of saying that for a given noise specification, the power distribution network cannot have more than some maximum impedance, and in fact the design of a power distribution system using frequency domain techniques takes this approach [14, 15].

For example, a distribution network having a 1-Ω impedance at a certain frequency will drop 1V for every amp of current drawn at that frequency. However, because the power distribution network includes inductors and capacitors, the impedance will not be constant for all frequencies; thus, the noise voltage will be different at each frequency.

A simple lumped RLC model as shown in Figure 6.16 is appropriate to model the interplane capacitance and decoupling capacitance at frequencies where the feature size is a small fraction of a wavelength. In this example, a 120-mil by 120-mil square section of a larger FR4 ($\varepsilon_r = 4.5$) PWB is modeled. The power and ground planes are separated by 5 mils.

The resistance, capacitance, and inductance of this section of PWB may be determined by (2.1), (3.4), and (4.15). The current flow is assumed to be across the length

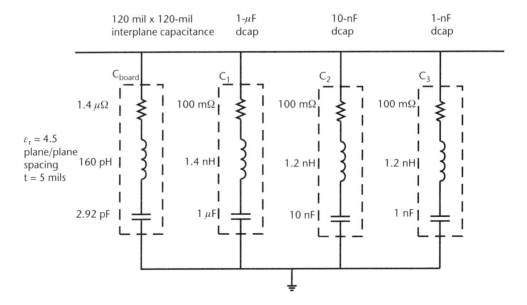

Figure 6.16 Power distribution RLC model.

of the cell (as opposed to flowing from corner to corner). The results are shown in the figure as C_{board}. The remaining three capacitors represent 1-F, 10-nF, and 1-nF decoupling capacitors added to the PWB. This model is for illustration only and has been simplified for clarity. Approximations for interconnect and via inductance are included as part of each capacitor model, but a proper model would include detailed circuits for vias and the decoupling capacitors and would properly account for resistance between elements. Chapter 10 discusses electrical modeling of decoupling capacitors.

For this model to be valid, the lumps must be much smaller than a wavelength. Using a factor of 10 as the definition of "much smaller," the segment of circuit board represented by C_{board} must be no larger than $\dfrac{\lambda}{10}$. Using this reasoning and applying (5.12) it's possible to derive the maximum frequency f_{max} for which a segment is valid:

$$f_{max} = \frac{11.8 \times 10^9}{10 \times length \times \sqrt{\varepsilon_r}} \tag{6.1}$$

where the term in the denominator is recognized as the speed of light in a vacuum given in inch-based units. In this example, the 120-mil segment is a valid lumped model for frequencies up to 4.6 GHz. A distributed model comprised of RLC or transmission line networks [19, 20] is required to properly analyze higher frequencies or larger circuit sizes. This is explored in Section 6.6.2.

Provided the frequency is below f_{max}, circuit theory may be used to analyze the circuit's response across various frequencies. For a review, see [21–23], but here we note that because the network of Figure 6.16 shows four different RLC circuits, there will be four separate series resonance frequencies alternating with parallel resonance frequencies. The lowest impedance value an RLC circuit will have occurs at the series resonance frequency, and at that frequency a simple RLC circuit will have a value equal to the value of the interconnect resistance. The impedance increases (appears inductive) for frequencies above series resonance. The parallel resonant frequency is the frequency of highest impedance for parallel combinations of RLC circuits. The resonate frequency for either a series or parallel resonant circuit is given by (6.2):

$$f_r = \frac{1}{2\pi\sqrt{LC}} \tag{6.2}$$

A Bode magnitude plot of the system shown in Figure 6.16 appears in Figure 6.17. The network's impedance is plotted against frequency on logarithmic scales. The solid curve is the network shown in Figure 6.16. The broken curve is the response of just the board itself (that is, the same network but with decoupling capacitors C1–C3 removed).

The bare board is seen to have smoothly decreasing impedance until it reaches its series resonance frequency at 7.4 GHz (which is above the validity of this model and so cannot be taken literally). Below this frequency, the circuit acts as a capacitor causing the impedance to decrease by a factor of 10 for every 10X increase in frequency.

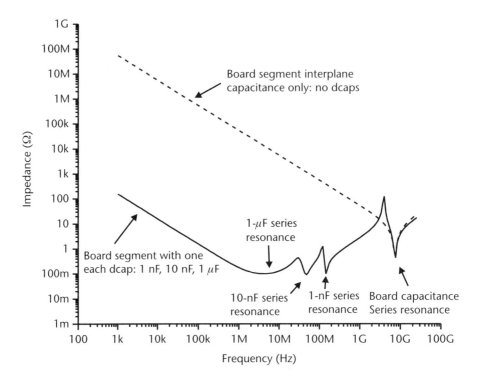

Figure 6.17   Bode magnitude plot of power distribution system as modeled in Figure 6.16.

As shown by the solid curve, the addition of decoupling capacitance to the bare board lowers the impedance to below that of just the bare board itself. However, the decoupled board is seen to have several changes in impedance, and it's possible for the impedance to be higher at a lower frequency than at a nearby higher one (namely, the resonate frequencies of the 10-nF and 1-nF capacitors). The impedance increases at parallel resonance, where the inductance of a decoupling capacitor resonates with the inductance and capacitance of the other capacitors present and is lower at the frequencies where the capacitor series resonates. For example, the 1-uF capacitor series resonates at about 1 MHz, and its inductance parallel resonates against the other capacitors at 13 MHz.

The peaks and valleys in impedance mean the noise voltage will not be the same for all frequencies. For example, assuming the same amount of current is drawn at all frequencies, the noise voltage will be nearly 10 times higher at 114 MHz (where the impedance is ~1.2Ω) than it will be at 143 MHZ (where the impedance is ~ 0.12Ω).

In practice, circuits do not draw the same current at all frequencies. Figure 6.18 shows the current drawn at various frequencies by the V_{dd} pin of an application-specific integrated circuit (ASIC) containing several serial transmitting devices operating at 3.125 Gbps as predicted by a SPICE model. The simulation included a package model but was connected to a perfect, lossless power supply so that the effects of board level power supply decoupling would not distort the natural behavior.

The figure was obtained by performing a fast Fourier transform (FFT) on the time-domain waveform predicted by SPICE. Mathematical analysis software such as

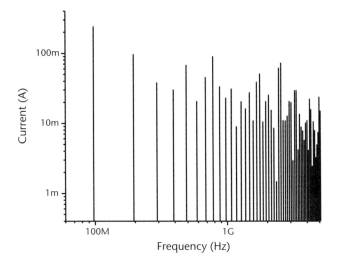

Figure 6.18   Current drawn by an ASIC transmitting several 3.125-Gbps data streams.

Mathcad [24] makes performing the FFT easy, or it can be performed within the SPICE simulator itself.

To obtain a proper FFT, the time steps used when setting up the SPICE simulation are critical, as that determines the displayed wave shape and so determines the frequency components contained in the waveform. Too course a time step can artificially remove high-frequency components and thus yield incorrect or misleading FFT results. The best results are obtained with a time step at least 50 to 100 times smaller than the fastest edge rate being measured.

The expected noise voltages at each discrete frequency shown in Figure 6.18 may be obtained by multiplying the current by the impedance predicted by the model in Figure 6.16. This is shown in Figure 6.19 for the frequency span of 100 MHz to 1.1 GHz.

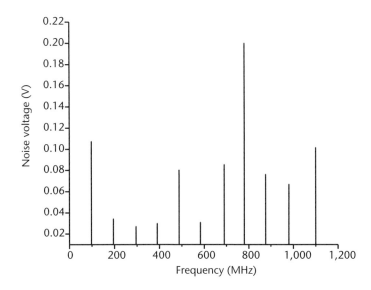

Figure 6.19   Noise voltage for Figure 6.16 given the noise current shown in Figure 6.18.

The analysis represented by Figures 6.17 and 6.18 is useful in showing the frequencies that need to be better attenuated and indicate the adjustments necessary to improve power supply integrity. For example, although the current at 98 MHz is more than 2.5 times as large as the current requirement at 782 MHz, the noise voltage at 98 MHz is seen to be only half as high. This is due to the decoupling scheme's efficacy at the lower frequencies. In this example, the greatest benefit will be obtained by adding capacitance to reduce the impedance at 782 MHz to below about 1Ω.

6.6.2 Distributed Power Supply Interconnect Model

The lumped model used in the previous discussion demonstrates the ideas behind using a frequency-domain approach to improve power supply decoupling, but the size of its lumps limits its upper frequency range. In fact, assuming 8b/10b nonreturn to zero (NRZ) signaling at 3.12 Gbps, the above model can't properly account for the behavior at frequencies beyond twice the fundamental frequency. As discussed in Chapter 7, such a waveform will have frequency components (harmonics) well above this frequency. Increasing the usable frequency requires segmenting the power and ground planes into numerous small cells, each sized to be a small fraction of the shortest wavelength (i.e., highest frequency) of interest. The segments may either be RLC PI networks or lossy transmission lines (so as to properly account for skin-effect losses) [20].

Such a model is shown in Figure 6.20, where a 20 × 20 grid of cells, each 0.030 in on a side, are arrayed. The dimensions of each 30-mil cell represent a tenth wavelength at 18.5 GHz, making the model valid to that frequency. As shown in the figure, an RLC PI network was created for each cell, and the cells were arrayed to represent a larger segment of the PWB. Increasing the array to 200 cells on a side would allow for the representation of a 6-in square area. Such a model would consist of 40,000 cells, for a total of 240,000 elements.

Even with all this complexity, the model only approximates the actual power system behavior, and the results should not be taken too literally. Nevertheless, this model is a useful tool that allows analyzing in a relative way the placement of sources, loads, and decoupling capacitances (complete with their own frequency-dependent RLC networks) in 0.03-in (30-mil) increments. Notice that in this model, the inductance in each cell is the same value (160 pH) as in Figure 6.16, but the capacitance is one-sixteenth the value. This is because the inductance is proportional to the cell's length-to-width ratio, while the capacitance is proportional to the cell's length times its width. The cell area of Figure 6.20 is one-sixteenth the area of the cell in Figure 6.16, and the capacitance is reduced proportionally. However, the cells are square and the current is assumed to flow across the length of the cells (rather than from corner to corner). This assumption makes the inductance identical for each cell, as the cell width-to-length ratio is equal to one in both cases. Of course from (4.15), changing the power to ground plate separation from the 5 mils assumed in this discussion would alter the 160-pH inductance per square cell (*per square*) figure.

The large number of elements and nodes makes the model unwieldy to run in SPICE-type simulators in the time domain (i.e., a transient analysis). Obtaining a

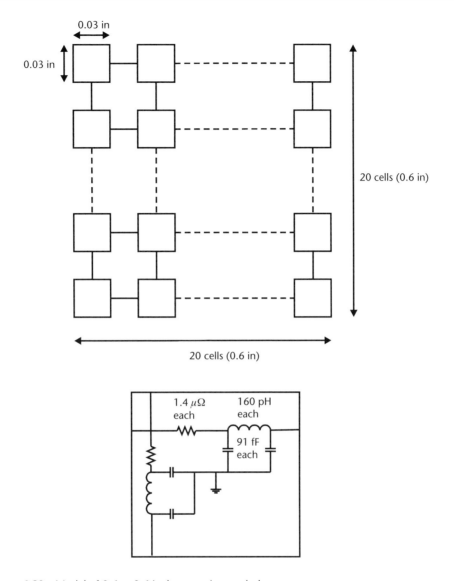

Figure 6.20   Model of 0.6 × 0.6 inch power/ground planes.

Bode plot by performing a frequency-domain analysis (an "AC" analysis in SPICE) is significantly quicker and, as previously demonstrated, gives deeper insight into the root causes of power supply noise.

The results of an AC-type SPICE analysis of Figure 6.20 appears in Figure 6.21. The three decoupling capacitors presented in Figure 6.16 are included. For this simulation, these capacitors were placed near the grid's center. Many more resonances appear than were present in Figure 6.16, and the impedances at the lower frequencies have benefited from the increased interplane capacitance (73 pF versus 2.9 pF). Relocating the decoupling capacitors would alter the frequency response and in a larger model would allow for capacitor placement analysis to be performed.

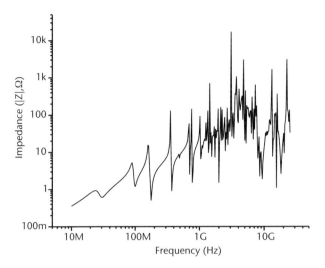

Figure 6.21 Bode magnitude plot of the model shown in Figure 6.20 with the three decoupling capacitors shown in Figure 6.16.

6.7 Connecting to Decoupling Capacitors

The way in which a decoupling capacitor is connected into the power system determines its parasitic *lead inductance* and thus is a factor in determining the capacitor's resonant frequency.

Often an integrated circuit's power and ground connections are made by connecting the pins by way of vias into power and ground planes running underneath the device. In this scheme, decoupling capacitors are located close by the integrated circuit and connect into these planes with their own vias. An alternative method is to connect the capacitors to the integrated circuit's pins with trace and then use vias to make the connection into the planes below. These schemes are illustrated in Figure 6.22.

The total loop inductance is twice the sum of the via's inductance plus the inductance due to any trace connecting the capacitors' mounting pads to the via. The inductance is two times this value in order to account for the inductance in both capacitors' leads. As shown in Figure 6.22(a), the trace inductance depends on the height h above the plane and its width w. A 5-mil-wide trace 5 mils above a return plane will have an inductance of about 10 pH/mil length. The results shown in the figure come from two-dimensional field-solving software, but the techniques described in Chapter 4 [the reciprocity principal, more directly from (4.16)] may be used to calculate this inductance for other situations.

6.7.1 Via Inductance

Equation (6.3) may be used to find the inductance of a via by itself, without accounting for any interconnecting trace [25]:

$$L_{via} = \frac{\mu_0}{2\pi}\left[h\ln\left(\frac{h + \sqrt{r^2 + h^2}}{r}\right) + \left(r - \sqrt{r^2 + h^2}\right)\right] \tag{6.3}$$

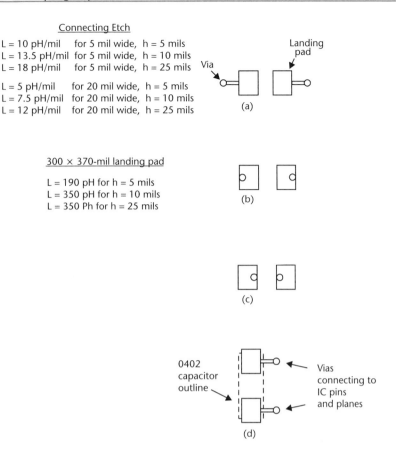

Connecting Etch

L = 10 pH/mil   for 5 mil wide, h = 5 mils
L = 13.5 pH/mil for 5 mil wide, h = 10 mils
L = 18 pH/mil   for 5 mil wide, h = 25 mils

L = 5 pH/mil   for 20 mil wide, h = 5 mils
L = 7.5 pH/mil   for 20 mil wide, h = 10 mils
L = 12 pH/mil   for 20 mil wide, h = 25 mils

300 × 370-mil landing pad

L = 190 pH for h = 5 mils
L = 350 pH for h = 10 mils
L = 350 Ph for h = 25 mils

Figure 6.22 Decoupling capacitor mounting: (a) using etch to connect via to pads, (b) placing vias in pads [see (6.3) for via inductance], (c) placing vias facing one another, and (d) connecting capacitor directly to IC pins.

where r is the via's radius and h is its length. The inductance is seen to decrease as the natural log of the via's radius, but increases with the via's length. This means that short, large-diameter vias will have lower inductance than long, small-diameter ones, and length is the bigger factor. Said another way, halving the via's length will cause a bigger percentage reduction in inductance than will doubling its diameter.

The following worked example shows how to use (6.3) and the data in Figure 6.22.

Example 6.1

Compute the loop inductance for configuration (a) shown in Figure 6.22. Include the via inductance, assuming the vias are 10 mils in diameter and the return plane is 10 mils below the boards surface. Assume an 0402 capacitor using a 300 × 370–mil pad and further assume the trace connecting the pad to the via is 5 mils wide and 50 mils long.

Solution

From (6.3) the via inductance L_{via} is 26.25 pH, and from Figure 6.22 the trace inductance is 10 pH/mil length × 50 mils = 500 pH. The pad itself has an inductance of

350 pH. The total loop inductance is therefore 2×876.25 pH $= 1.75$ nH. This is nearly three times larger than the capacitor's equivalent series inductance (ESL) (see Figure 10.4). The loop inductance more than doubles to 3.7 nH if the planes are moved to 25 mils below the surface.

This example assumes that the two vias connecting the capacitors are far apart and so coupling is not beneficial. Removing the trace by placing the vias in the pads [as shown in Figure 6.22(b)] eliminates the sizable trace inductance. Doing that as well as locating the vias next to each other to take advantage of mutual inductance between the vias as shown in Figure 6.22(c) can nearly cut the via inductance in half [26]. Another assumption is that the capacitor is placed at the edges of the mounting pad so that the current must flow across the entire face. In fact, the capacitor will be more centered on the pads, thus reducing the length of pad metal that will carry current to the via. An additional simplification in this example is that the inductance of the planes is not accounted for. This can easily be included and should be done when comparing this form of decoupling to the style illustrated in Figure 6.22(d), where the capacitor is connected to the integrated circuit's pins by trace before it connects to the power and ground planes.

6.8   Summary

The relationship between a signal conductor and its return fundamentally determines the transmission line's electrical characteristics. Proper referencing of all signals helps to suppress EMI and is necessary for best signal integrity. This is true when passing differential or single-ended signals along PWB trace, through a connector, or when crossing a mote. Using diff pairs to cross a mote usually does not result in adequate high-speed signal integrity.

Special attention is necessary to ensure that signals are referenced to the proper power plane in split-plane situations or when using a power plane to form a stripline. Inductance in the power system effectively blocks high-frequency return currents from flowing between the power supplies back to the source. It's best (but in some applications not practical) to form stripline with two ground planes rather than one ground and a power plane.

The current drawn by integrated circuits is not uniform across all frequencies. High-frequency harmonics are usually present. Integrated circuits requiring large switching currents often have decoupling capacitance located on chip and on package, effectively integrating the current draw over frequency. This somewhat reduces the magnitude of the high-frequency currents that must be supplied by the PWB power system.

The power system inductance and capacitance cause the power system to have an impedance that changes with frequency. Resonances will occur at many different frequencies as the various capacitances' and inductances' series and parallel resonate with one another. High impedance occurs at parallel resonance while the lowest occurs at series resonance. Any current component occurring at one of the parallel resonance frequencies will create large noise voltages at that frequency.

The frequency-dependent impedance of the power system has direct and indirect effects on signal quality.

The direct effect is for the power-system impedance to hinder the high-frequency return currents from completing their paths back to the source. This causes the high-frequency components of the waveform to become attenuated (or, in severe cases, completely removed), distorting the driven signal as measured at the load.

The indirect effect is for the noise voltage present on the power system to hinder or prevent transmitting devices from driving the highest frequency components. This results from *voltage sag* or *droop* and is caused by the power system's impedance dropping various amounts of voltage at each frequency component making up the switching current waveform. Excessive power supply noise can also increase transmitter jitter from an integrated circuit.

Crosstalk will increase for signals changing layers, as the vias will often preferentially couple to each other rather than to a reference plane. This can be minimized by separating vias spatially or by adding a return path via along with each signal via.

The signal-to-return ratio of signals passing through a connector must be carefully managed. Adequate ground connections must be provided through the connector to ensure the ground planes of the two cards being interconnected are at the same potential. Even when signaling differentially, an inadequate number or placement of ground connections through the connector will compromise signal quality and will lead to EMI problems.

The size of antipads and use of nonfunctional pads must be carefully considered in dense pin fields. Wide trace reduces resistive losses but often is too wide to route through fine-pitch pin fields, especially when routing edge-coupled differential pairs. The change in impedance as a signal travels along a via (such as a connector pin or layer-changing via) will increase for thicker PWBs, and coupling from the pin to any planes will be the highest for small antipads. Large antipads reduce available routing channels and can lead to ground starvation. Detailed simulation or measurement of test articles is the best way to make the trade-off between trace width, pad/antipad size, and board thickness.

References

[1]   Hubing, T. H, et al., "Power Bus Decoupling on Multilayer Printed Circuit Boards," *IEEE Trans. Electromagnetic Compatibility*, Vol. 37, No. 2, May 1995, pp. 155–166.

[2]   Johnson, H. W., and M. Graham, *High-Speed Digital Design*, Englewood Cliffs, NJ: PTR Prentice Hall, 1993.

[3]   Chen, N., "Modeling of Nonideal Return Paths on Multilayer Package," *IEEE 2002 Electronic Components and Technology Conference*, San Diego, CA, May 28–31, 2002, pp. 57–61.

[4]   Tektronix Inc, "Using TDR to Help Solve Signal Integrity Issues," Applications Note, 2001, pp. 1–16, www.tektronix.com.

[5]   Liaw, H., and H. Merkelo, "Crossing the Planes at High Speed," *IEEE Circuits and Devices Magazine*, Vol. 13, No. 6, November 1997, pp. 22–26.

[6]   Fornberg, P., et al., "The Impact of a Nonideal Return Path on Differential Signal Integrity," *IEEE Trans. Electromagnetic Compatibility*, Vol. 44, No. 1, February 2002, pp. 671–676.

[7]   Chen, J. et al., "Power Bus Isolation Using Power Islands in Printed Circuit Boards," *IEEE Trans. On Electromagnetic Compatibility*, Vol. 44, No. 2, May 2002, pp. 373–380.

[8]   Norman, A. J., et al., "Experimental and Simulation Analysis of Single and Differential Signals Changing Layers," *IEEE International Symposium on EMC*, Vol. 2, Montreal, Quebec, August 13–17, 2001, pp. 1088–1091.

[9]   Xiaoning, Wei Cui, et al., "EMI Resulting from Signal Via Transitions Through the DC Power Bus," *IEEE International Symposium on EMC*, Vol. 2, Washington, D.C., August 21–25, 2000, pp. 821–826.

[10] O'Sullivan, C., and N. Lee, "Ground Starvation Effects on Multi-Layer PCBs," *IEEE International Symposium on EMC*, Vol. 2, Washington, D.C., August 21–25, 2000, pp. 113–116.

[11] Ye, X., et al., "High-Performance Inter-PCB Connectors: Analysis of EMI Characteristics," *IEEE Trans. On Electromagnetic Compatibility*, Vol. 44, No. 1, February 2002, pp. 165–174.

[12] Montrose, M. I., *EMC and the Printed Circuit Board*, New York: IEEE Press, 1999.

[13] Drewniak, J. L., et al., "Modeling Power Bus Decoupling on Multilayer Printed Circuit Boards," *IEEE International Symposium on EMC*, Chicago, IL, August 22–26, 1994, pp. 456–461.

[14] Smith, L. D., et al., "Power Distribution System Design Methodology and Capacitor Selection for Modern CMOS Technology," *IEEE Trans. On Advanced Packaging*; Vol. 23, No. 3, August 1999, pp. 284–291.

[15] Ricchiuti, V., "Power-Supply Decoupling on Fully Populated High-Speed Digital PCBs," *IEEE Trans. Electromagnetic Compatibility*, Vol. 43, No. 4, November 2001, pp. 671–676.

[16] Fan, J., et al., "Quantifying SMT Decoupling Capacitor Placement in DC Power-Bus Design for Multilayer PCBs," *IEEE Trans. Electromagnetic Compatibility*, Vol. 43, No. 4, November 2001, pp. 588–599.

[17] Young, B., *Digital Signal Integrity*, Chapter 4, Englewood Cliffs, NJ: PTR Prentice Hall, 2001.

[18] Madou, A., and L. Martens, "Electrical Behavior of Decoupling Capacitors Embedded in Multilayered PCBs," *IEEE Trans on Electromagnetic Compatibility*, Vol. 43, No. 4, November 2001, pp. 549–555.

[19] O'Sullivan, C. B., et al., "Developing a Decoupling Mythology with SPICE for Multilayer Printed Circuit Boards," *IEEE International Symposium on EMC*, Denver, CO, August 9–14, 1998, pp. 652–655.

[20] Keummyung, L., and A. Barber, "Modeling and Analysis of Multichip Module Power Supply Planes," *IEEE Trans on Components, Packaging and Technology, Part B*, Vol. 18, No. 4, November 1995, pp. 628–639.

[21] Johnson. D. E., et al., *Basic Electric Circuit Analysis*, 5th Ed., Chapter 14, Englewood Cliffs, NJ: PTR Prentice Hall, 1995.

[22] Van Valkenburg, M. E., *Analog Filter Design*, New York: Holt, Rinehart and Winston, 1982, pp. 68–76.

[23] Skilling, H. H., *Electrical Engineering Circuits*, 2nd Ed., Chapter 18, New York: John Wiley & Sons, 1965.

[24] Mathsoft Engineering & Education, Inc., Cambridge, MA.

[25] Goldfarb, M. E., and R. A. Pucel, "Modeling Via Hole Grounds in Microstrip," *IEEE Microwave Guided Letters*, Vol. 1, No. 6, June 1991, pp. 135–137.

[26] Tang, G., "Surface Mount Capacitor Loop Inductance Calculation and Minimization," *IEEE International Symposium on Electromagnetic Compatibility*, Aug. 24–28, 1998, pp. 505–510.

CHAPTER 7

Serial Communication, Loss, and Equalization

7.1 Introduction

Parallel bus signaling is ubiquitous on PWBs, and although loss effects are often not discussed, high-performance parallel bus signaling (such as source synchronous signaling) is amply described in the literature (for example, see [1–4]). On the other hand, baseband serial transmission (where a serial data stream is transmitted without modulation) is not as well covered in the signal integrity literature. This type of signaling is becoming increasingly popular as integrated circuit technology permits the production of reliable and repeatable serial transmitters and receivers capable of multigigabit-per-second data rates. This technology is widely used for signaling across long backplanes and cables and is now becoming mainstream enough to gain acceptance as a way to signal locally between ASICs collocated on a PWB.

In this chapter, we discuss baseband serial signaling, with an emphasis on the effects loss has on signal quality. Although this chapter explicitly focuses on serial transmission, the discussion of harmonics and distortion is equally relevant to the lines within a parallel data bus.

The chapter begins with the briefest possible review of Fourier analysis and shows the relationship between spectral content and pulse characteristics. Line and block codes are next discussed in the context of the frequency content of a data stream, and ISI (the bane of high-speed serial or parallel signaling) is introduced. Eye diagrams as a diagnostic tool naturally follow. Equalization and preemphasis (essential for serial transmission at gigabit-per-second rates) are examined next, which leads into the use of ac coupling between a serial transmitter and receiver. The chapter completes with an analysis on the proper way to calculate the capacitor's value.

7.2 Harmonic Contents of a Data Stream

The elements of a repetitive stream of rectangular pulses are defined in Figure 7.1.

The pulse transitions from 0 to A volts with a rise and fall time t_r. It's τ sec wide at its base, and t_w sec wide at its top. It has a period of T seconds.

The ratio of the pulse's width to the repetition period is the waveforms *duty cycle* (δ), as defined in (7.1):

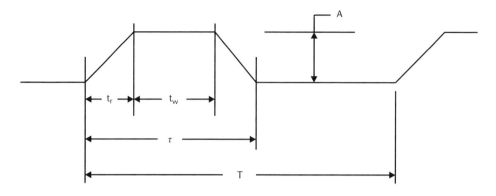

Figure 7.1 Pulse characteristics definition.

$$\delta = \frac{\tau}{T} \tag{7.1}$$

The duty cycle is usually given in percentages [i.e., a square wave has a 50% duty cycle because $T = 2(t_r + t_w)$] and its width is measured at specific voltage levels (e.g., 50% points). The waveform is measured at the base in Figure 7.1.

From Fourier analysis, it's known that any periodic waveform can be reconstructed by adding a series of sinusoids with the proper amplitude and phase characteristics. This series—called the Fourier series—may be written as the summation of sins, cosines, a combination of the two, or in exponential form. Depending on the characteristics of the waveform, the series may need an infinite number of terms.

The Fourier series of a repeating waveform is given in trigometric form as a series of sine and cosines in (7.2):

$$f(t) = \frac{A_o}{2} + \sum_{n=1}^{\infty} \left(a_n \cos n\omega_o t + b_n \sin n\omega_o t \right) \tag{7.2}$$

The A_o term is the dc component (the signal's mean value), while the a_n and b_n terms are the *Fourier coefficients* for each harmonic ($n\omega_o$). The *fundamental radian frequency* ω_o is given in (7.3):

$$\omega_o = 2\pi f_o \tag{7.3}$$

The waveform's period determines the fundamental frequency f_o (7.4):

$$f_0 = \frac{1}{T} \tag{7.4}$$

From (7.2), each harmonic is separated in frequency by an amount equal to f_o. This is shown formally in (7.5):

$$f_n = n f_o \tag{7.5}$$

where *n* is the harmonic number.

For example, a square wave with a mid amplitude point swinging symmetrically about zero and starting at $t = 0$ will have all of the a terms equal to zero and the b terms equal to zero for even values of n [5]. The Fourier series therefore contains only the odd harmonics (7.6):

$$f(t) = \frac{4A}{\pi}\left(\sin \omega_0 t + \frac{1}{3}\sin 3\omega_0 t + \frac{1}{5}\sin \omega_0 t + \ldots\right) \tag{7.6}$$

where A is the square wave's peak amplitude. The Fourier coefficients C_n are $C_1 = 1$, $C_3 = \dfrac{1}{3}$, $C_5 = \dfrac{1}{5}$, and so on.

7.2.1   Line Spectra

The amplitude of each frequency in the Fourier series (the Fourier coefficient, C_n) may be individually plotted as an amplitude line spectrum (presented briefly in Chapter 6), or just the envelope encompassing the amplitude peaks may be plotted. For those waveforms having only two phase values (generally 0° and 180°), the phase may be shown on the same plot as the amplitude by setting the polarity of the spectral lines. In other cases, a separate phase plot is used to display the phase spectrum.

The line spectra for a train of very sharp-edged 1-V, 500-ps-wide pulses having a 1-ns period is shown in Figure 7.2.

The magnitude of each of the first 20 harmonics is shown in Figure 7.2 as positive values. This is similar to spectrum analyzers, which generally only display the coefficient's magnitude and not its phase. Often the frequency is plotted on a logarithmic scale, but because the frequency span displayed is not too great, this was not done in Figure 7.2. Additionally, using a linear scale in the figure improves its

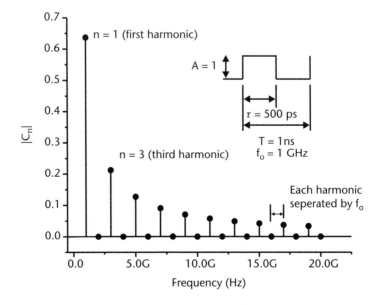

Figure 7.2   Line spectra of a 500-ps-wide pulse with a 1-ns period.

clarity, as it avoids crowding the graph at the upper frequencies where the distance between divisions would be small on a logarithmic scale.

In agreement with (7.4) and (7.5), the harmonics are seen to occur every 1 GHz, and consistent with (7.6) only the odd harmonics have nonzero values.

7.2.2 Combining Harmonics to Create a Pulse

An example of how a pulse can be produced by adding together sinusoids is shown in Figure 7.3. A 1-ns square wave is constructed by summing the first nine odd harmonics.

The waveform is seen to become more pulselike as higher frequency harmonics are added in. Combining just the first seven odd harmonics (f_1, through f_7) results in a waveform with noticeable ripple in the pulse's top portion but distantly sharper edges than that of the fundamental sine wave. The pulse shape improves as additional harmonics are added. As shown, including the next odd harmonic (f_9) produces an adequate waveform with only a small amount of ripple. It actually has the same amplitude as the other waveform, but its been scaled smaller in Figure 7.3 to make it more visible.

Looking at Figure 7.3 in the reverse way, as a subtractive process rather than an additive one, gives initial insight as to how the lowpass characteristics of a lossy transmission line distorts a pulse. If a 1-ns square wave is sent down a transmission line that passes the first nine harmonics without altering their amplitudes or phase relationship, then it will appear formed as shown in Figure 7.3. However, the edges become more gradual and the pulse more rounded as successive harmonics are removed. Although it illustrates the effect on a square wave as harmonics are removed, this example is somewhat contrived because lossy transmission lines do not abruptly remove harmonics in this way, the phase relationship

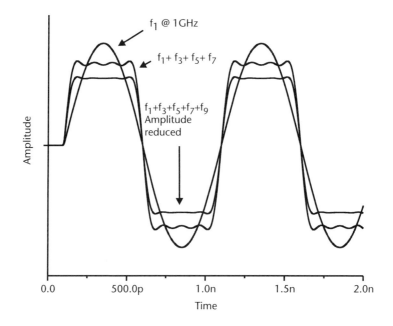

Figure 7.3   Adding harmonics to create a 1-ns square wave.

between the harmonics does not remain constant (as tacitly assumed in this example), and the type of data streams we're concerned with in this book (called NRZ data, as described in Section 7.3) do not contain only square waves. Instead, as we'll see subsequently, the data stream contains pulses of various widths and the transmission line gradually, rather than abruptly, attenuates and phase shifts the harmonics.

Combining the same harmonics used to create Figure 7.3 but altering their amplitudes and phase relationship yields very different waveforms, as shown in Figure 7.4.

The topmost solid curve is the identical waveform appearing in Figure 7.3, which is constructed by adding the first nine odd harmonics. The dotted curve shows the effects of phase distortion, where the phase relationship between each harmonic is intentionally misadjusted. The resulting curve is no longer flat topped and shows peaking as the improper phasing causes the sinusoids to combine inappropriately. The dashed curve shows the effects of amplitude distortion, where each harmonic is reduced in amplitude by various amounts but the phase relationship remains correct. Although flat topped, the pulse is seen to be rounded: the improper combination of amplitudes has removed the pulse's sharp edges, resulting in a smoothed pulse having lower overall amplitude. The lower solid curve shows the effects of both amplitude and phase distortion. The pulse is reduced in amplitude, has rounded edges, and has sloping top and bottom portions. Most significantly, the base of this pulse has spread outside of the region occupied by the undistorted pulse. We'll return to this phenomenon in Section 7.6.

Clearly, a transmission line that does not preserve the proper amplitude and phase relationships between the harmonics will yield distorted pulses, and because line losses are proportional to line length the distortion will grow with the transmission line's length.

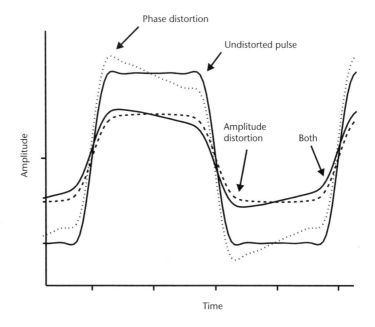

Figure 7.4   Effects of altering harmonic's amplitude and phase.

7.2.3 The Fourier Integral

The Fourier series may be used to obtain line spectra of a periodic signal, but the more general Fourier integral is used to obtain the frequency components of a stream of various width pulses or of a single pulse. Knowledge of a pulse stream's spectral content is useful, as it determines the channel's necessary bandwidth. This analysis is available in signals and systems or linear circuits texts (see [5, 6]), and only the results are presented here.

The Fourier coefficients C_n for a rectangular waveform with perfectly vertical edges ($t_r = 0$) such as that displayed in Figure 7.2 is given in (7.7) [7, 8]:

$$C_n = Aav \frac{\sin\left(\pi \frac{n\tau}{T}\right)}{\pi \frac{n\tau}{T}} \tag{7.7}$$

where Aav is the signals average value, as given in (7.8):

$$Aav = 2A\frac{\tau}{T} = 2A\delta \tag{7.8}$$

The results from (7.7) appear in Figure 7.5, plotted as a line spectrum for a pulse width $\tau = 500$ ps and $T = 4$ ns.

Harmonics are spaced every $\frac{1}{T} = 250$ MHz, and have a zero value every $\frac{1}{\tau} = 2$ GHz. In comparison to the results shown in Figure 7.2, when $T = 1$ ns, increasing the period to 4 ns results in the harmonics being more closely spaced (250 MHz versus 1 GHz) and the amplitude being reduced [as required by (7.5)]. In both cases, zero crossings still occur every 2 GHz, as τ has not changed.

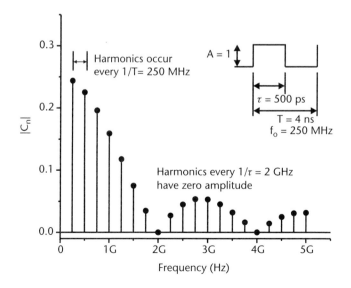

Figure 7.5 Line spectra for $\tau = 500$ ps and $T = 4$ ns.

Continuing to increase T in this way further reduces the spacing between harmonics, and ultimately if T is made large enough, individual harmonics can no longer be observed. Instead of individual spectral lines, the Fourier integral may be used to obtain a density function (the pulse's *spectral density, G(f),* units of volts per Hertz) whose area shows the how a range of frequencies contribute to the waveforms makeup (see [5, 6, 9]) (7.9):

$$G(f) = V\tau \frac{\sin \pi \dfrac{\tau}{T}}{\pi \dfrac{\tau}{T}} = V\tau \frac{\sin \pi\tau f}{\pi\tau f} \tag{7.9}$$

In general, $G(f)$ will have a complex value for each frequency. The amplitude distribution is found from the function's magnitude; the phase spectrum is found from its argument.

Equation (7.9) is of the form $\dfrac{\sin(\pi x)}{\pi x}$, which is known as the *sinc* function. The absolute value of this function is plotted in Figure 7.6 for the specific case where $\tau = 500$ ps and $T = 4$ ns.

The sinc function has a value of zero for integer multiples of π, which corresponds to frequencies of $\dfrac{1}{\tau}$ Hz [6, 9], which in turn corresponds to the zero crossings in Figure 7.5.

7.2.4   Rectangular Pulses with Nonzero Rise Times

Practical data pulses require time to transition between logic levels. This makes the rise time $t_r > 0$, and (7.4) does not strictly apply.

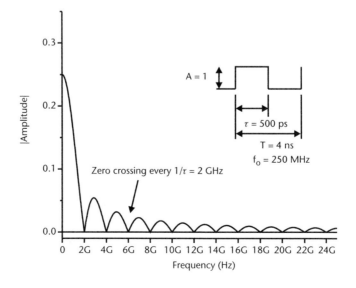

Figure 7.6   Pulse train spectral envelope.

The Fourier coefficients for a rectangular waveform with nonzero rise times is given in (7.10) [7, 10]:

$$C_n = 2Aav \left(\frac{\sin\left(\pi n \dfrac{t_r}{T}\right)}{\pi n \dfrac{t_r}{T}} \right) \left[\frac{\sin\left(\pi n \dfrac{t_r + t_w}{T}\right)}{\pi n \dfrac{t_r + t_w}{T}} \right] \tag{7.10}$$

where Aav is the pulses average amplitude as given by (7.11):

$$Aav = A \frac{t_r + t_w}{T} \tag{7.11}$$

Equation (7.10) is useful in determining the line spectrum of a repeating waveform, such as a series of clock pulses. Many high-performance oscilloscopes have a spectrum analyzer function where they perform an FFT on the waveform displayed in the time domain. Equation (7.10) may be used as a reference to help determine beforehand the results from these instruments.

The line spectra of a 500-ps-wide pulse (at its base) having a 4-ns period with zero and 75-ps rise times is shown in Figure 7.7.

The filled circles reproduces the data in Figure 7.5 for $t_r = 0$. The hollow circles show the effects when the rise time is changed to 75 ps.

Notice that the line spectrum for the $t_r = 75$ ps case has mostly lower values as compared to the pulse with zero rise time, and that the zero crossings (the absence of a particular harmonic) occurs at different frequencies. This demonstrates the advantage of employing integrated circuits that incorporate edge rate control because by retarding the driven edges, such integrated circuits reduce the amplitudes of the Fourier coefficients.

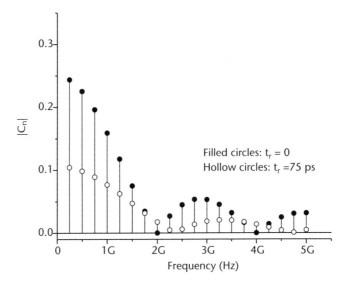

Figure 7.7   Pulse with $\tau = 500$ ps, $T = 4$ ns for $t_r = 0$ and 75 ps.

7.3   Line Codes

The previous sections have assumed a binary-type signaling, where a logic "1" is represented by a voltage being present for a fixed amount of time, while a logic "0" is the absence of a voltage. In fact, there are many ways to electrically represent the logic states on the wire. The various schemes are called *line codes*, with perhaps the one most familiar to designers accustomed to parallel signaling being the unipolar NRZ format shown in the top portion of Figure 7.8. In this context *unipolar* means a signal switching between 0 and V volts, where V is usually (but not necessarily) a positive voltage. This differs from *bipolar* signaling, where the signal swings between $-V$ and $+V$ (e.g., RS232 or RS485 signaling). In that case, the code would be called a bipolar NRZ line code. Many other line codes have been devised to suit specific applications (see [6]).

The line code shown is called an NRZ code because it remains at V for the duration of the bit time. This contrasts with return to zero (RZ) signaling (shown at the drawing's bottom), where the pulse representing a logic 1 is at V for only part of the bit time (usually half a bit time) before it returns to zero.

Designers most familiar with parallel bus signaling at CMOS logic levels [e.g., high-speed transceiver logic (HSTL) or stub series terminated logic (SSTL)] sometimes mistakenly refer to the top waveform in Figure 7.8 as an RZ code. The confusion apparently arises because the signal swings from V to zero volts, but doesn't go below ground to a negative voltage to represent a logic zero. As we've seen, such a signal would properly be called a bipolar signal.

It's not necessary for the signal to take on only one of two states. While such binary signaling is probably the most familiar, multilevel signaling (sometimes called *m-ary* signaling, with *m* representing the number of levels) can be used to increase the data transfer rate without a corresponding increase in channel bandwidth [8]. This type of signaling is also called pulse amplitude modulation (PAM) and is becoming increasingly popular for signaling at multigigabit rates [6, 8, 11–14].

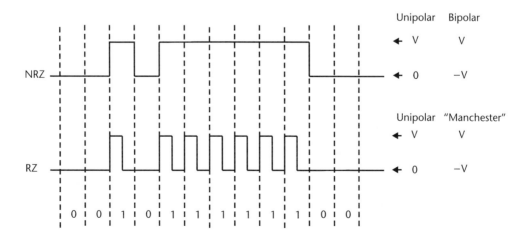

Figure 7.8   RZ and NRZ line codes.

7.4   Bit Rate and Data Rate

A series of sharp-edged, flat-topped pulses is shown in Figure 7.9. This data stream represents an alternating 1/0 NRZ data pattern and is idealized because the edges are sharp and the pulses stay within predefined sampling windows. As indicated earlier and discussed in Chapter 5, in actual systems transmission losses cause the pulses to become rounded and to smear outside of their assigned time slots. This is further examined in Section 7.6.

The pulse's width τ is the *bit time* (sometimes called the *bit cell time* or simply the *cell time*) and in the figure is shown to be 400 ps. This is a basic measurement of time in serial signaling and is often referred to as a *unit interval* (UI). In this example, the waveform's period T consists of two UI, or 800 ps. From (7.4), this yields a fundamental frequency f_o for this data pattern of $\dfrac{1}{800 \text{ ps}} = 1.25$ GHz.

The *data rate* is defined as the number of bits transmitted per interval of time and has units of bits per second (bps), (7.12):

$$\text{data rate} = \frac{\text{number of bits transmitted}}{\text{time interval}} = \frac{1}{UI} \text{ bps} \qquad (7.12)$$

In an alternating 1/0 pattern as appears in Figure 7.9, two bits are transmitted every two UI. In this example 1 UI = 400 ps, and from (7.12) this yields a data

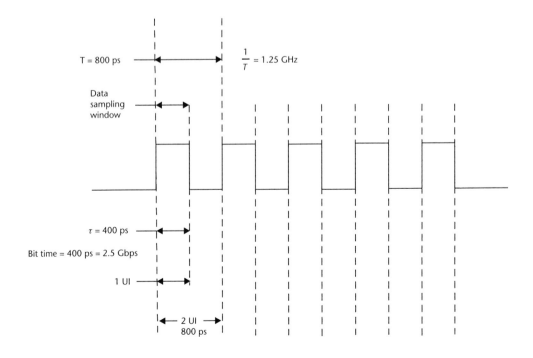

Figure 7.9   Idealized alternating NRZ 1/0 data stream at 2.5 Gbps.

rate of 2.5 Gbps. Notice that in this example, the frequency is half the signaling rate in bps.

Because the UI rather than T is often used when discussing serial transmission, it's convenient to have a way to directly convert between the two. As shown in Figure 7.9, the smallest period T occurs in a data stream having an alternating 1/0 pattern in which case $T = 2UI$. It follows that the highest fundamental frequency for a binary NRZ data stream is therefore (7.13):

$$f_{o\_max} = \frac{1}{2UI} \qquad (7.13)$$

It's worth noting here that, as discussed subsequently, the signal will contain harmonic frequencies higher than f_{o_max}.

Other patterns of data will result in lower fundamental frequencies. For example, the data rate in Figure 7.10 is still 2.5 Gbps, but a string of three back-to-back 1s has reduced f_o to 625 MHz during the sampling period shown.

In fact, over a suitable sampling window, f_o of an unencoded NRZ data stream can range from dc up to f_{o_max} as given in (7.13), unless some intentional mechanism raises the lowest possible frequency by restricting the number of back-to-back same-polarity bits. Block codes such as the 8b/10b code described in the next section provide this mechanism, effectively placing a lower limit on the frequency content

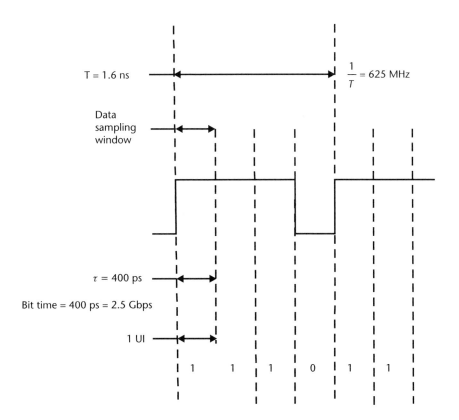

Figure 7.10   NRZ data stream showing frequency reduction caused by a string of same polarity bits.

of the data stream. This has the advantage of restricting the frequency range over which the receiver's clock recovery circuits (usually a phase locked loop or a delay locked loop) must operate yet remain locked. It also provides a limit on the frequency range over which the communication channel's amplitude characteristic must remain flat and the phase characteristic must increase linearly (as discussed in Chapter 5). It also makes dc blocking between the transmitter and receiver practical. The blocking can be done with transformers, but for cost and area conservation reasons it's more common to use capacitors at very high data rates when many serial paths' areas are integrated onto one circuit board. Selecting these capacitors is discussed in Section 7.10.

7.5 Block Codes Used in Serial Transmission

Before it can be transmitted serially, the parallel data is formatted into a serial data stream by a device commonly called a serializer/deserializer (SERDES). This serialized data is then serially transmitted by the SERDES, where it's received by another SERDES that deserializes the data stream back into a parallel format. The transmit and receive SERDES have separate clocks (called the *bit clock* or *reference clock*), which are frequency multiplied by an on-chip phase locked loop (PLL). The data stream is driven out and received at a rate determined by the internal PLLs. The PLL multiplication factor is often programmable and so allows the SERDES to transmit at various bit rates. This can be very helpful during system debug, as it allows a link to be run at a slower speed to initially perform protocol and logic tests. A typical connection between SERDES is shown in Figure 7.11.

The parallel data sent to the transmitting SERDES in Figure 7.11 is not merely latched into a shift register and shifted out serially, bit for bit. Instead, it's desirable to frame data into *blocks* having defined signaling and frequency content characteristics. To accomplish this the parallel port data (the *parallel word* or the *input word*) is encoded into a *character* having a larger number of bits. For example, a 4-bit input word block might be encoded into a 5-bit character for serial transmission.

There are many *block codes* (see [9]), but one that's commonly used for serially transmitting binary data over copper and fiber optic cables is the 8b/10b code [15]. In this code, an 8-bit parallel word is encoded into 10 bits for serial transmission. The receiving SERDES uses the block code rules to recover the original 8 bits from the 10 transmitted. Because 10 bits can encode four times as many characters as can be created by the 8-bit input word (1,024 versus 256), the encoding has space for additional characters and also has enough room to allow certain bit sequences to be

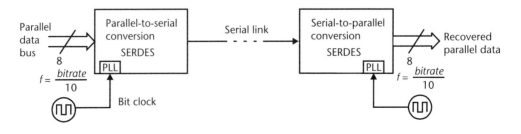

Figure 7.11 Generalized SERDES serial connection.

excluded. For example, a 10-bit string of all same-sense bits would not be allowed and so would be excluded from the data space and could not be transmitted.

Certain special sequences of bits created by the encoding are called *control characters* to distinguish them from *data characters*. These are also sometimes referred to as control and data *symbols*, but we'll refrain from using this terminology so as to better distinguish between single characters and groups of characters, which we'll call symbols.

Twelve control characters are defined by the 8b/10b encoding [15]. Data characters are usually identified as *Dn.n*, and control characters as *Kn.n*, where *n* is a decimal digit.

Block codes are described by their *run length*, *disparity*, and *digital sum variation*. The run length is the maximum number of same polarity bits appearing back to back in the data stream, while the disparity is a measure of the difference between the number of 1s and 0s within a given portion of the data stream (generally defined as a single data block). The disparity is (+) if there are more 1s than 0s, (−) if there are more 0s than 1s, and zero if the number of 1s and 0s are equal. The digital sum variation (DSV) is similar to (and sometimes confused with) disparity, but where disparity concentrates on the imbalance of 1s and 0s within a data block, the DSV is the running total of the number of 1s and 0s in the data stream. It's updated continuously and is useful in detecting bit errors in the data stream [9].

For an 8b/10b code, the DSV is six, the maximum run length is five, and the disparity is either 0 or ±2, meaning that a correctly formed 10b word may have five 0s (and so five 1s), or six 0s (and four 1s), or four 0s (and six 1s) [15]. No other possibilities are permitted.

A small portion of an 8b/10b coding table is given in Table 7.1 [16].

As an example, a *D0.0* character (all eight bits in the input word set to zero) would not be serially transmitted as a string of 10 0s back to back. Instead, from Table 7.1, the bit stream would either be 100111 0100 or 011000 1011, depending on the value of the running disparity. The first stream would be selected if the previous block of data had more 0s than 1s (disparity−); the second stream would be selected for disparity+. In either case in this example, the stream transmitted is *disparity neutral* (disparity = 0), as the word contains the same number of 1s as 0s.

Not all code groups are disparity neutral. Notice that the *D21.7* character has an unequal number of 1s and 0s. Sending back-to-back *D21.7* characters therefore

Table 7.1   8b/10b Coding Table Fragment

Input Word Value (Hexadecimal)	Input Word Binary Value (MSB/LSB)	Byte Name	Encoding Disparity−	Encoding Disparity+
00	000 00000	D0.0	100111 0100	011000 1011
01	000 00001	D0.1	011101 0100	100010 1011
F5	111 10101	D21.7	101010 1110	101010 0001
1C	000 11100	D28.0	001110 1011	001110 0100
4A	010 01010	D10.2	010101 0101	010101 0101
FC	111 11100	D28.7	001110 1110	001110 0001
BC	000 11100	K28.5	001111 1010	110000 0101
FC	111 11100	K28.7	001111 1000	110000 0111

results in two different 10-bit words being serially transmitted: assuming the disparity is initially negative (more 0s than 1s), a 101010 1110 will be sent for the first $D21.7$. Because that sequence is disparity+, the second $D21.7$ will be transmitted as 101010 0001. Taken together, this string of 20 bits is disparity neutral, with a maximum run length of four (corresponding to the string of four 0s appearing in the second $D21.7$).

The control characters such as K28.5 are used to identify byte boundaries and thus allow the data stream to be separated into definable packets for parsing and decoding. Other control characters are used to specially identify certain packets or to adjust timing. These details are beyond the scope of this book, but knowledge of control character characteristics is important to those performing signal integrity analysis because these characters tend to have the highest concentration of back-to-back same-polarity bits. As we'll subsequently see, this means a string of control characters will have a lower frequency content than a string of data characters. This is particularly true of a class of control characters called *comma* characters. In Table 7.1, the K28.5 and K28.7 characters are comma characters. By forcing the appropriate data and control characters, it's possible during debug to observe the serial link's behavior at different frequencies. This is useful in determining the degree of operational margin due to ISI (discussed in the next section) and is particularly helpful when performing interoperability testing between assembled equipment.

7.6 ISI

In earlier chapters we've seen how a lossy transmission line attenuates and phase distorts sinusoids. Signal degradation will occur unless the line can properly maintain the amplitude and phase relationships between all the harmonics being transmitted. This is rarely the case when signaling on PWB traces and cables. The effect a bandwidth-limited channel has on signal propagation is well covered in most signals and systems texts (for example, see [17–19]). In this and succeeding sections, we'll accept that, as discussed in Chapter 5, lossy transmission lines act as a lowpass filter, bandwidth limiting the communication channel and thereby altering the amplitude and phase relationship between the waveforms harmonics. In the time domain, a pulse traveling through a bandwidth-limited channel will experience spreading, and if severe enough it will smear into neighboring time slots. This causes ISI—the distortion of a data bit within a symbol due to interference caused by one or more earlier data bits, either from that symbol or a preceding one. Another cause of ISI is residual energy left in the line by reflections from an impedance discontinuity or mismatch somewhere along the transmission line. Another type of distortion is *dispersion*, which arises when the harmonics making up a pulse arrive at the receiver at different times.

7.6.1 Dispersion

The harmonics composing a pulse do not all travel down a transmission line at the same velocity. This phenomenon is called *dispersion* [20, 21], and it causes each of the harmonics forming a pulse to arrive at the load at different times. From

Chapter 5, a delay can be equated to a shift in phase [see (5.10)], so harmonics arriving at different times are reaching the receiver out of phase with one another. Said another way, the lossy transmission line has altered the original amplitude and phase relationship of the sinusoids forming the pulse.

As was shown in Figure 7.4, a pulse undergoing amplitude and phase distortion will become smaller, become rounded, and its base will widen. If the widening is severe enough, energy will smear into the next bit cell time to alter the characteristics of the bit present there.

This is demonstrated in Figure 7.12, which shows the progressively worse distortion experienced by a rectangular 400-ps-wide 1-V pulse having a 100-ps rise time as it travels down a 1m-long lossy stripline.

The pulse is observed every 0.25m along its length. The pulse is seen to shrink in amplitude and to grow in width as the pulse propagates down the line. To illustrate this, the original pulse is shown as a broken curve, placed to outline the position the bit would have if the line were lossless. At the $l = 0.25$m mark, some of the pulse's energy is dispersing into the next bit cell time, and progressively more dispersion is evident as the pulse continues to traverse the line. A good portion of the bit's falling edge is outside its allocated time slot at a distance of $l = 1$m. Any pulse appearing in the next bit cell would be compromised.

7.6.2 Lone 1-Bit Pattern

In Figure 7.13, a pulse stream containing a 1111010010 bit pattern is sent down the same line that's illustrated in Figure 7.12. As before, the launched bits in this stream are rectangular 400-ps-wide 1-V pulses with a 100-ps rise time.

In this case the line is sampled at two points ($l = 0.25$m and $l = 1$m). It's evident that the string of back-to-back 1s reaches the receiver with higher amplitude than does the single (or *lone*) 1 pulse. When measured at $l = 0.25$m the wide, low

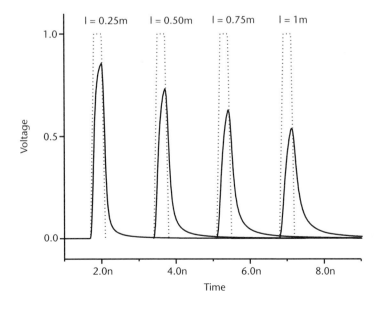

Figure 7.12 400-ps, 1-V pulse propagating down a 1m-long, 5-mil-wide lossy stripline on FR4.

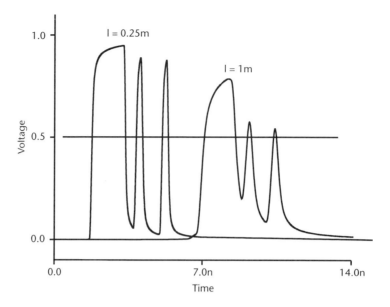

Figure 7.13   Bit stream traveling down a lossy line.

frequency pulse consisting of four back-to-back 1s very nearly reaches 1V, while the single 1s have lower amplitude. The $l = 1$m length shows that this data-dependent reduction in amplitude increases with longer length lines. At 300 mV, the lone 1's amplitudes are sufficient for proper reception, but because it has been offset by the low-frequency pulse, the high-frequency pulse just barely cross the 500-mV threshold level [22, 23]. In fact, patterns of pulses can be devised to explicitly exacerbate amplitude or timing uncertainties. The uncertainty in a pulses timing is called *jitter* and has several components as described in [24].

7.7   Eye Diagrams

The small bit cell times but the large number of bits comprising a symbol make it difficult to analyze the amplitude and time characteristics of a serial data stream such as that shown in Figure 7.14.

Individual bits can be examined in detail, but it's not easy to determine if a particular bit somewhere in the data stream has inadequate timing or amplitude margin. The number of bits in a symbol and the number of possible combinations of symbols constituting a data packet makes examination of each individual bit impractical.

Eye diagrams (so called because the clear portion vaguely resembles an eye shape) are created by overlaying the positive and negative going pulses present in the data stream. By synchronizing the horizontal sweep with the data pattern and using an *infinite persistence* display, the oscilloscope or *communication analyzer* (a sampling oscilloscope specifically designed to record and display serial bit streams) will show all transitions superimposed. All of the bits are therefore displayed simultaneously, and any data dependency will be evident [25, 26]. Some CAD simulation tools can produce eye diagrams (for example, see [27–29]).

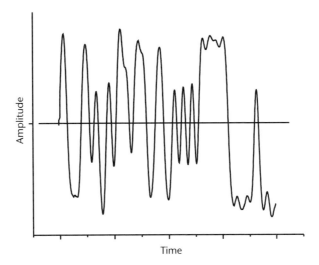

Figure 7.14   Serial data stream.

If all of the pulses have the same timing and amplitude characteristics, all of the rising edges will be placed directly on top of one another, as will all of the falling edges. Figure 7.15 shows the resulting eye diagram.

If the eye diagram is measured at the receiver, it's easy to determine the receiver's operating margin by comparing the received eye to a *mask* illustrating the receiver's operating limits [25].

The rising and falling edges won't all be perfectly aligned in actual systems, and as we've seen the pulses won't all have the same amplitudes. Instead, imperfections in the transmitter, the transmission path, and characteristics of the data pattern itself will cause the edges to transition at different times and will prevent the amplitude of each pulse from being the same as that of preceding ones. A mask outlining the region of proper operation of an imperfect data stream is shown in Figure 7.16.

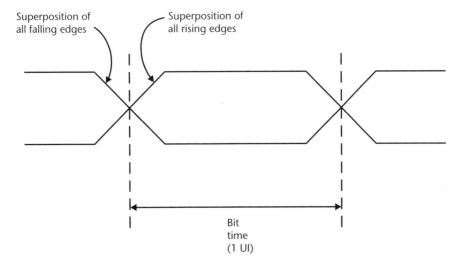

Figure 7.15   Eye diagram with ideal, perfectly timed pulses.

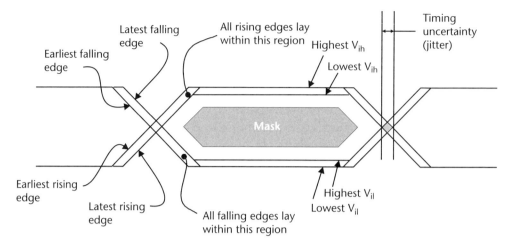

Figure 7.16 Imperfect pulses encroaching on a data mask.

The receiver is expected to properly operate if the waveforms do not extend into the region outlined by the mask. In fact, any transgressions into the mask region should cause the receiver to misinterpret the data. Although such events are undesirable and indicate a degraded or flawed transmission path, occasional transgressions into the mask region (*mask hits*) are not necessarily fatal to the communications link. If the data eye is generally clear and only occasionally experiences a mask hit, then the error-correction mechanisms inherent to the chosen block code will often be capable of correcting the error.

Jitter is visible by the width of the region, where the high and low going traces cross. This can also be displayed as a histogram on some communication analyzers. Amplitude uncertainty (caused by baseline wander or ISI) appears as a vertical reduction in the eye opening. The display of a communication analyzer uses color (or grayscale, as reproduced in this book) to show the frequency of occurrences within a specific region.

7.8   Equalization and Preemphasis

Equalization networks are designed to have a frequency response that is approximately the inverse of the transmission line, thus making the frequency response at the receiver uniform (*equalizing* the frequency response) across all frequencies of interest [18].

The correction mechanism may be incorporated digitally within the transmitter (where it's commonly called *preemphasis*), at the receiving end of the transmission line (where it's usually simply called *equalization*), or in both places.

7.8.1   Preemphasis

Preemphasis at the transmitter is a routine feature of gigabit SERDES (for example, see [30–33]). In these designs, the driver's output current is increased over the nominal value when the bit is transitioning [23]. The typical preemphasis algorithm may examine a single adjacent bit or may examine multiple bits in the data stream [34]. In the single bit case, the current is reduced by a predetermined amount only when the bit changes state. Back-to-back same-polarity bits are driven with a lower, nominal current value. The high-frequency transitions in the data pattern (such as a 1 appearing after a long series of 0s) are thus *preemphasized* relative to the back-to-back same-polarity bits. The result is that the waveforms high-frequency portion is transmitted with more energy than the lower frequency portions. A typical example is shown in Figure 7.17.

The solid line shows the transmitter's output voltage with preemphasis off: each bit and each sequence of bits are driven with the same current, making the output voltage the same for each type of transition.

As shown by the dotted curve, the pulse shapes are quite a bit different when preemphasis is activated. Three pulses are shown as they are exiting the transmitter: the first is a string of three back-to-back 1s, the second a string of two back-to-back pulses, and finally a single width pulse. The initial string of three back-to-back 1s has a very different characteristic than does the single 1 at the end of the stream. We see that the output is initially driven high but then throttles back to a lower plateau value when driving the remaining portion of the pulse. The initial transition has been emphasized over the other 1s in the pulse. The initial transition of the second pulse has also been emphasized and so shows similar amplitude and decay characteristics. Because the pulse width is not long enough, the amplitude does not plateau to the same voltage as the wider first pulse. The lone 1 at the end of the data stream is driven with an amplitude equal to the initial value given the other pulses but is never throttled back.

Notice that an alternating 1/0 string would always be preemphasized and so would be transmitted with more energy than a pattern containing lower frequency components such as control characters. Referring to Table 7.1, arraigning the SERDES to transmit back-to-back D10.2 characters would cause pulses to be

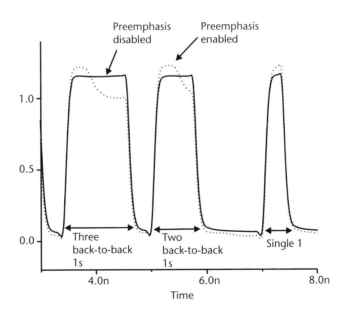

Figure 7.17   Transmitter output waveform without and with preemphasis.

continually preemphasized, while K28.5 characters show a mix of behaviors, including a *lone 1* characteristic. These test patterns are useful when examining the channel's frequency response, especially if the link's response can be observed with the preemphasis switched on and off.

The improvement preemphasis can make on a signal is quite remarkable. The waveform received at the end of a 9-m (29.5-ft) long 50-Ω 28-AWG shielded twisted pair cable is shown in Figure 7.18. A pseudorandom bit stream (PRBS) pattern of valid 8b/10b characters is being transmitted at a data rate of 2.5 Gbps (400-ps UI), and the transmitter preemphasis is turned off. Although an eye is visible, the signal is so degraded as to be unusable.

The received eye is greatly improved when transmitter preemphasis is activated, as Figure 7.19 shows.

The eye appearing in Figure 7.19 is recorded with the same scale as that used in Figure 7.18. The improvement is evident and is sufficient to allow proper link operation.

The loss characteristics of a specific application will generally not be known beforehand to the SERDES integrated circuit designer, and often a SERDES is designed to properly preemphasize a long, lossy transmission line. Such a channel is perceived to be the most challenging, and successfully driving an extremely long line gives a competitive marketing advantage. This means the SERDES preemphasis characteristics have been tuned to operate best in that one, specific high-loss environment and will perform suboptimally anywhere else. In fact, unless the SERDES provides the ability for the end user to adjust the precompensation levels, it's quite possible for a SERDES to operate better when transmitting along long traces or cables than it will for short ones.

As a practical matter, the end user usually experimentally determines the appropriate equalization settings for each channel during product simulation and debug. The values are then hard wired or loaded via software during a power-up sequence. A difficulty with this scheme is that the equalization setting is static, and there is no

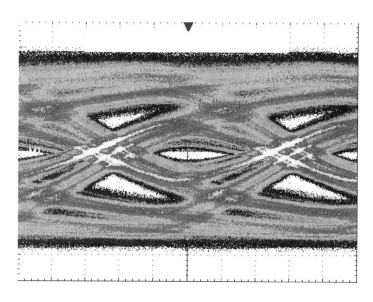

Figure 7.18   Received waveform with single bit preemphasis turned off.

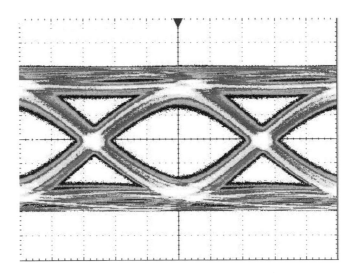

Figure 7.19   Improvement with preemphasis turned on.

feedback mechanism in the transmitter to dynamically adjust the coefficients to account for aging or changing environmental factors. More sophisticated systems employ training sequences to automatically tune the transmitter's compensation settings for optimum performance and periodically make adjustments as environmental conditions warrant. The transmitter/receiver pair are said to automatically *adapt* themselves to the changing conditions of the communications channel. This *adaptive equalization* scheme requires a matched transmitter/receiver pair that must communicate with each other (generally by way of a low-performance *sideband* link) in a known and predictable way [14, 23]. Presently there are no common standards between manufactures for the sideband protocol. This makes interoperability between different manufacturers impractical.

7.8.2   Passive Equalizers

Passive equalizers are filter circuits placed in series with a transmission line to correct for frequency-dependent amplitude or phase distortion. A typical application with cables is the use of a highpass filter to attenuate the lower frequency harmonics, bringing their amplitudes more into line with the heavily attenuated higher frequency ones.

Synthesis of passive equalizers is beyond the scope of the book, but it's noted that many circuit forms using resistance, inductance, and capacitance are possible. An example of a simple resistor and capacitor (RC) equalizer is given in Figure 7.20 and in the next section. Those wishing to explore the many other topologies are referred to [17, 19] for more details.

In gigabit signaling, the equalizer may be incorporated as an intrinsic part of the integrated circuit receiver, or when signaling over cables the equalizer can be included in with the cable assembly itself (generally in the connector portion of the cable). In either case, the goal is to flatten the frequency response of the transmission path so that the harmonics combine at the receiver in such a way as to properly reconstruct the transmitted pulse.

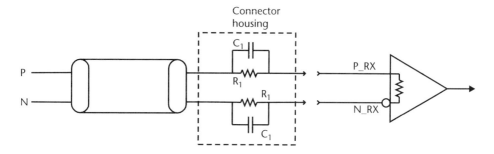

Figure 7.20   RC equalizer circuit.

A simple highpass filter circuit sometimes used to equalize cables is shown in Figure 7.20. The RC network is present on both the P and N sides of a differential pair and usually appears in the connector at the receiving end of the cable (although it would work just as well if placed at the transmitting end). Because the circuit is so simple, it's often mounted on a small circuit board located inside the connector shell. Surface mount resistors and capacitors are used, thereby saving space and limiting parasitic effects.

The effectiveness of passive equalizers is demonstrated in Figures 7.21 and 7.22. The received eye at the end of an unequalized 1m (∼ 3.3 ft) long, 50-Ω 24-AWG shielded twisted pair cable is shown first in Figure 7.21. The transmitters preemphasis is turned off, and it's sending the same PRBS pattern at 2.5 Gbps as was used to create Figures 7.18 and 7.19. Notice that for these figures, the vertical scale is about 20% larger.

Although in Figure 7.21 the eye opening is large enough for proper reception, the waveform is jittery and greatly benefits from equalization, as Figure 7.22 shows.

The addition of a simple passive equalizer of the type shown in Figure 7.19 has significantly improved the received data eye. The amplitude of the eye's clear portion has nearly doubled and the jitter noticeably reduced.

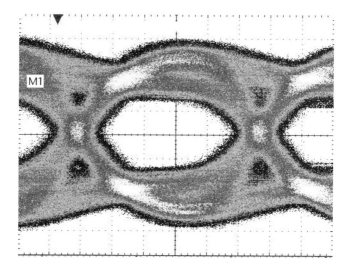

Figure 7.21   1m-long cable with no equalization.

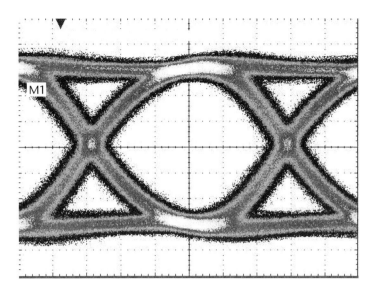

Figure 7.22   1m-long cable with equalization.

7.8.3   Passive RC Equalizer

Referring to the circuit schematic in Figure 7.19, the values for resistor R1 and capacitor C1 are presented in [35], and although reformatted they are offered here without proof in (7.14) and (7.15).

$$R_1 = Z_o (K - 1) \tag{7.14}$$

$$C_1 = \frac{\dfrac{\sqrt{K}}{K - 1}}{2\pi f_c Z_o} \tag{7.15}$$

where Z_o is the transmission line's characteristic impedance and f_c is the equalizer's desired highpass cut-off frequency. The factor K determines the filter's insertion loss (the amount of attenuation the filter provides, measured at the cutoff frequency [36]) and the degree of flatness in the filter's transfer curve. In practice, typical values of K range from just over one to less than five. For 50-Ω trace and cable systems with $f_c \sim 1.25$ GHz, C_1 typically has values in the sub 10s of pF range and R_1 in the many tens of ohms.

The insertion loss at a frequency f given in decibels as (7.16) [35]:

$$IL(f) = 10 \log \left(1 + \frac{K^2 - 1}{1 + K \left(\dfrac{f}{f_c} \right)^2} \right) \tag{7.16}$$

Although higher values of K produce a flatter frequency response, it's seen from (7.16) that increasing K causes the insertion loss to increase.

An example of the way in which K and f_o interact appears in Figure 7.23. The frequency response of a 9.5-in-long, 5-mil-wide, 50-Ω lossy stripline on FR4 that has not been equalized appears in the top curve. At 1.25 GHz, the signal strength has been reduced by roughly 25% over its value at 200 MHz. The remaining three curves show the effects when measured at P_RX and N_RX of adding a simple RC equalizer, as depicted in Figure 7.19. The bottommost curve shows an amplitude change of less than 10% over the same frequency range. Clearly the equalizer has compensated for the lossy line's frequency behavior, making the combination much flatter across frequency. This flatness comes with a cost: although the harmonics are all attenuated by about the same amount (especially for higher values of K), the entire signal has been reduced in amplitude. Lower values of K have less insertion loss but the frequency response is not as flat. As shown, a practical equalizer of this type trades off the insertion loss for flatness.

7.9   DC-Blocking Capacitors

Two SERDES are connected together by a long transmission line, as shown in Figure 7.24. Capacitors C_p and C_n provide dc isolation between the transmitter and receiver, allowing the receiver to respond only to the ac portion of the waveform without regard to the bias level.

Besides improving noise margin by allowing for local rebiasing at the receiver, the capacitors (commonly called *dc-blocking* or *ac-coupling* capacitors) permit *hot swap* operation, whereby the receiver is powered down while the transmitter continues to operate. The capacitors prevent dc from flowing from the transmitter through electrostatic discharge (ESD) structures or other parasitic elements in the powered-off receiver. These parasitics are described in [34]. An active transmitter driving into a powered-off receiver is a common occurrence in large network switches or other

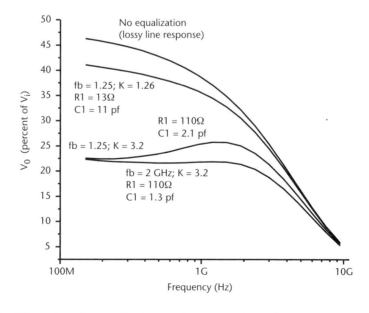

Figure 7.23   Effects of passive equalization on a lossy transmission line.

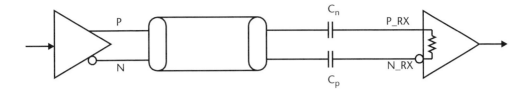

Figure 7.24   Use of dc-blocking capacitors on a serial link.

equipment where the receiver is on a separate card from the transmitter. Fault conditions make it possible for the receiver to be unpowered for long periods while the transmitter continues to operate. Under these circumstances, the series capacitors prevent large continuous currents from being sourced by the transmitter into essentially the short circuit represented by the receiver parasitic diodes.

The capacitors can be placed at either the transmitting or receiving end of the line and still provide dc isolation. However, in practice they are nearly always placed at the receiver.

A common mistake is to select too small a value for the coupling capacitor. A small capacitance will cause the signal at the receiver to be distorted, resulting in a change in the reference level (the *baseline*) as a function of the data pattern's duty cycle. This effect (*baseline wander*) results in reduced operating margin.

Conversely, a capacitance too large will require a long time to establish an appropriate steady-state bias across the capacitor. Often the time to initially establish the proper bias is not of practical concern because the transmitter will usually send synchronization codes (*idle characters* or *training sequences*) essentially indefinitely until the receiver signals that it is properly detecting the data stream. This implies that a proper bias has been established. Once this occurs, the actual data is sent and if the capacitance is large enough relative to the maximum run length, the bias point will not move very much.

The capacitor's frequency response, especially the way in which the capacitor's value is affected by frequency, temperature, and voltage is another factor to consider when selecting the coupling capacitor. In Chapter 10, these things are discussed in detail, but here it's noted that large-valued capacitors suggest physically large capacitors, and this implies high ESL (essentially a parasitic inductance in series with the capacitor). It's desirable to keep ESL low because it forms a lowpass filter with the landing pads shunt parasitic capacitance.

One final point about the capacitor's physical size concerns its ESR. This is a resistance modeled in series with the capacitance and inductance of the capacitor, and its value is very dependent on package size and capacitance value. A model showing ESL and ESR plus the frequency response is presented in Chapter 10 (Figures 10.2 and 10.3). In general, those packages that are longer than they are wide will have the highest ESR and should be avoided. Good choices for package sizes (*body styles*) are 0508 (which is wider than long, so has low ESR and ESL, see [37]), 0402, and 0603. Capacitors in the 0603 body style will often (but not always) have a lower ESR than those in the 0402 package, but generally a higher ESL. In many applications, X7R is adequate when selecting the dielectric for the coupling capacitor C_c, especially if its nominal value is many times larger than the minimum value required. Calculation of C_c is shown in the following section.

7.9.1 Calculating the Coupling Capacitor Value

Calculating the minimum capacitance value can be done using either a time-domain or frequency-domain approach. The frequency domain approach is presented here. A time-domain analysis is presented in [9].

The circuit representation of Figure 7.24 is presented in Figure 7.25.

The coupling capacitor C_c and the termination resistor R_{term} form a highpass filter. From filter theory (for example, see [36]), the circuit's *cut-off frequency* f_c is:

$$f_c = \frac{1}{2\pi R_{term} C_c} \tag{7.17}$$

The cutoff frequency is the frequency where the filter's output voltage is $\frac{1}{\sqrt{2}} = 0.707$ times the input voltage. This is illustrated in Figure 7.26, which shows the transfer function of an RC highpass filter.

Because it's a first-order filer, the transition from the stopband to the passband is somewhat gradual, changing at a rate of 20 dB for every tenfold increase in

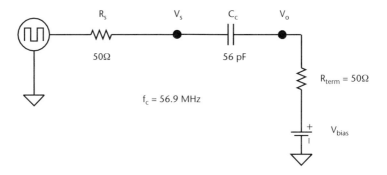

Figure 7.25 Coupling capacitor circuit representation.

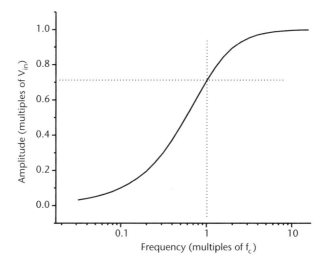

Figure 7.26 RC highpass filter output characteristics.

frequency. The output reaches 0.707 times the input voltage at a frequency equal to f_c. In practical terms, a frequency of 20 times f_c or more is passed through the filter without much loss (at that frequency, the output amplitude is 0.9988 time the input amplitude).

When using (7.17) to calculate C_c, a mistake to avoid is to assume that the data stream's lowest frequency is f_{o_max} [given in (7.13)]. In fact, as we've seen, the lowest frequency can be quite a bit below this and is determined by the characteristics of the block code. The significance of this is illustrated in Example 7.1.

Example 7.1

Calculate the coupling capacitor value for the 8b/10b 2.5-Gbps NRZ data stream shown in Figure 7.27. Assume the line impedance is 50.

Solution

As shown in Figure 7.25, a 50-Ω source drives the transmission line, which is ac coupled by the coupling capacitor C_c to a differential amplifier receiver. In this example, the receiver is perfectly terminated into 50Ω, and for simplicity has no parasitic capacitance. Likewise, the electrical effects of the receiver's micropackage are not included, nor are the capacitor's ESR or ESL.

The data stream used to stimulate the Figure 7.25 network is a K28.7 comma character followed by a D10.2 data character, which is shown in Figure 7.27 to have a lowest frequency occurring at 277.8 MHz.

It will first be erroneously assumed that the lowest frequency present is f_{o_max} = 1.25 GHz.

The highpass filter's cut-off frequency should be much lower than the lowest frequency present to ensure the signal's fundamental is not overly attenuated. Assuming the lowest frequency is 1.25 GHz and designing for a factor of 20 times

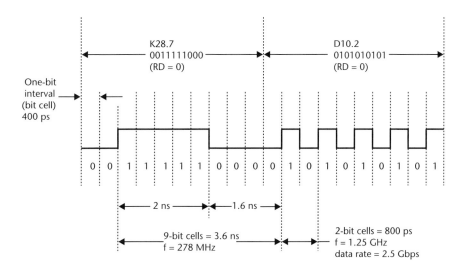

Figure 7.27   Data stream showing two characters each with running disparity = 0.

less than that makes the desired cut-off frequency f_c = 62.5 MHz. Solving (7.13) for C_c yields a minimum capacitance of 51 pF. A stock 56-pF capacitor will be selected for the coupling capacitor C_c, yielding an f_c of 56.9 MHz.

The effects in using a capacitor with too small a value are evident in the bottom portion of Figure 7.28, which shows the waveform measured at the receiver when coupled by a 56-pF capacitor.

The tops and bottoms of the received signal droop, and the signal does not symmetrically swing about the 0.5-V centerline as expected. The center of the bit cell and its mid-voltage point is shown with an X in the graph. Ideally, the cell center should fall directly on top of the 0.5-V horizontal line, as that is the receiver's trigger point. For best noise margin, the receiver's trigger would need to be moved as shown by the X in each bit cell. It's clear that the large duty cycle caused by the string of back-to-back 1s has charged capacitor C_c to a voltage different from what it had been when an alternating 1/0 pattern (50% duty cycle) was received. The result is a shift downwards in the midpoint that if large enough will exceed the receiver's operational range.

The top portion of Figure 7.28 shows the received waveform when C_c is calculated using 278 MHz. Again assuming f_c is 20 times less and solving (7.13) for C_c yields C_{min} = 229 pF. A stock value of 270 pF is chosen. The bottom graph shows significant improvement, but some baseline wander is still evident.

However, as shown in Figure 7.29, increasing the capacitor to 2,700 pF results in essentially no baseline wander and a waveform with no droop. In all portions of the waveform, the cell center properly lines up with the mid-rail voltage (0.5V). To obtain this performance, the cut-off frequency f_c has been lowered to 1.2 MHz (more than a factor of 200 below the 278 MHz first chosen). This is practical at the data rate used in the example, as the capacitor value is not overly large and can be obtained in a small package having low ESR and ESL.

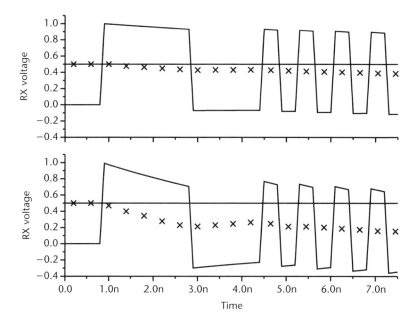

Figure 7.28   Baseline wander with C_c = 56 pF (bottom) and 270 pF (top).

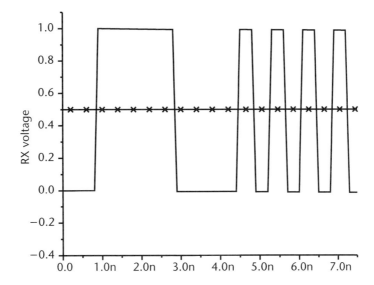

Figure 7.29   Received voltage with C_c = 2,700 pF.

7.10   Summary

The fundamental frequency f_o is determined by the period of the pulse train. Harmonics are integer multiples of f_o. A sharp-edged pulse (one with a small t_r) of a given period will have harmonics at the same frequencies as a pulse having a poorer rise time with the same period. However, the sharper pulse's upper harmonics will have a greater magnitude. Said differently, the harmonics of the pulse with the sharper edge rate will take longer to fade to insignificance.

Receiver equalization is a way to flatten the frequency response of a lossy transmission line, and if properly adjusted for a particular network, it can significantly improve signal quality.

A lossy line experiencing both phase and amplitude distortion will undergo a reduction in pulse height and dispersion, where the pulse smears outside of its assigned bit boundary. Low-frequency energy present in a transmission line will interfere with the more attenuated high-frequency energy (the lone 1 phenomenon), there by corrupting data by causing ISI.

Line codes (such as NRZ) define the way in which information is represented on a line. Block codes (such as 8b/10b) define the way in which pulses are grouped into characters.

The lowest frequency possible on a channel is determined by the characteristics of a particular block code, while the highest frequency is determined by the narrowest pulse width.

Transmitter precompensation is very common in gigabit SERDES and is effective at compensating for the loss of high-frequency harmonics. Precompensation is passive in that, once set, it does not automatically adjust for changing environmental conditions. Adaptive equalization automatically changes the transmitter's characteristics as needed to obtain optimum channel response.

When ac coupling, a serial transmitter and receiver set the receivers cut-off frequency f_c to be at least 20 times lower that the lowest frequency. The capacitor's ESL and ESR are important considerations when selecting the capacitor. Use mica, X7R, or C0G ceramic in a 0508 or 0402 SMT body style.

References

[1] Takasaki, Y., *Digital Transmission Design and Jitter Analysis,* Norwood, MA: Artech

[2] Nadolny, J., and M. Kravets, "Active Cable Assemblies for 10 Gigabit Ethernet," *Design-Con 2003,* Santa Clara, CA, January 27–30, 2003.

[3] Stefanos, S., et al., "High-Speed Inter-Chip Signaling," in *Design of High-Performance Microprocessor Circuits,* A. Chandrakasan, W. Bowhill, and F. Fox, (eds.), New York: IEEE Press, 2001, pp. 397–42

[4] Johnson, H. W., and M. A. Graham, *High-Speed Digital Design,* Englewood Cliffs, NJ: PTR Prentice-Hall, 1993.

[5] Lauterbach, M.,"Getting More out of Eye Diagrams," *IEEE Spectrum,* Vol. 34, No. 3, March 199

[6] Couch, Leon W. II, *Digital and Analog Communications Systems,* 5th Ed., Englewood Cliffs, NJ: Prentice-Hall, 1997.

[7] Van Valkenburg, M. E., *Analog Filter Design,* New York: Holt, Rienhart and Winston, 1982.

[8] Bateman, A., *Digital Communications,* Reading MA: Addison-Wesley, 1998.

[9] Bissell, C. C., and D. A. Chapman, *Digital Signal Transmission,* Cambridge, England: Cambridge University Press, 1996.

[10] Montrose, M., *EMC and the Printed Circuit Board,* New York: IEEE Press, 1999.

[11] Farjad-Rad, R., et al., "A 0.3m CMOS 8 Gb/s 4-PAM Serial Link Transceiver," *Journal of Solid State Circuits,* Vol. 35, No. 5, May 2000, pp. 757–764.

[12] Zerbe, J., et al., "Equalization and Clock Recovery for a 2.5–10 Gb/s 2-PAM/4-PAM Backplane Transceiver Cell," Paper 4.6, *International Solid State Circuits Conference,* Santa Clara, CA, February 9–13, 2003.

[13] Zerbe, J., et al., "1.6 Gb/s/pin 4-PAM Signaling and Circuits for a Multi-Drop Bus," *IEEE 2000 Symp. On VLSI Circuits,* Honolulu, HI, June 15–17, 2000, pp. 128–131 .

[14] Sonntag, J., et al., "An Adaptive PAM-4 5 Gb/s Backplane Transceiver in 0.25um CMOS," *IEEE 2002 Custom Integrated Circuits Conference,* Orlando FL, May 12–15, 2002.

[15] Widmer, A. X., and P. A. Franaszek, "A DC-Balanced, Partitioned-Block, 8B/10B Transmission Code," *IBM J. Res. Development,* Vol. 27, No. 5, September 1983, pp. 440–451.

[16] Infiniband Trade Association,"Infiniband Architecture Specification, Vol. 2, Release 1.1," November 2002.

[17] Proakis, J. G., *Digital Communications,* New York: McGraw Hill, 2000.

[18] Horowitz, M., et al., "High-Speed Electrical Signaling: Overview and Limitations," *IEEE Micro,* Vol. 18, No. 1, January/February 1998, pp. 12–24.

[19] Federal Telephone and Radio Corp., *Reference Data for Radio Engineers,* 3rd Ed., New Y

[20] Matick, R., *Transmission Lines for Digital and Communication Networks,* New York: IEEE Press, 1969.

[21] Miner, G., *Lines and Electromagnetic Fields for Engineers,* New York: Oxford University Press, 1996.

[22] Dally, W. J., and J. W. Poulton, "Transmitter Equalization for 4-Gbps Signaling," *IEEE Micro,* Vol. 17, No. 1, January/February 1997.

[23] Dally, W. J., and J. W. Poulton, *Digital Systems Engineering,* New York: Cambridge University Press, 1998.

[24] Haykin, S., and B. Van Veen, *Signals and Systems,* 2nd Ed., New York: John Wiley & Sons, 2002.

[25] Takasaki, Y., *Digital Transmission Design and Jitter Analysis,* Norwood, MA: Artech

[26] "LVDS Signal Quality: Jitter Measurements Using Eye Patterns Test Repost #1," Application Note 977, National Semiconductor, October 1994.

[27] SiAUDITOR® Software, SISOFT Inc., www.sisoft.com.

[28] Iconnect® TDR Software, TDA Systems, www.tdasystems.com.

[29] HSPICE® Software, Synopsys, Inc., www.synopsys.com.

[30] Texas Instruments Corp., "TLK3104SA Quad 3.125Gbps Serial Transceiver Data Sheet, Revision 2.2," March 5, 2001.

[31] Vitesse Semiconductor Corp., "VSC7226-01 Double-Speed Multi-Gigabit Interconnect Chip Data Sheet, Revision 2.6," October 24, 2001.

[32] Marvelle Corp., "High Performance Backplane Design Using the Marvell Alaska X Quad 3.125 BG/s SERDES," Applications Note, November 2001.

[33] Burns, D., et al," Design Techniques for High-Speed Source Synchronous Busses," *DesignCon2002*, Santa Clara, CA January 28–31, 2002.

[34] Nadolny, J., and M. Kravets, "Active Cable Assemblies for 10 Gigabit Ethernet," *DesignCon 2003*, Santa Clara, CA, January 27–30, 2003.

[35] Terman, F. E., *Radio Engineers Handbook*; New York: McGraw Hill, 1943.

[36] Lauterbach, M.,"Getting More out of Eye Diagrams," *IEEE Spectrum*, Vol. 34, No. 3, March 19

[37] "Low Inductance Capacitors 0612/0508/0306 LICC," Data Sheet, AVX Corp.

CHAPTER 8

Single-Ended and Differential Signaling and Crosstalk

8.1 Introduction

The focus of previous chapters has been on the characteristics of single lines, with only casual mention of coupled lines. This chapter discusses the way transmission lines couple to each other and so induce noise voltages to other signal lines, and the effect they have on one another's impedance and timing.

Two side-by-side microstrips (traces "E1" and "E2") are shown in Figure 8.1. If the lines are close enough, the magnetic and electric fields will encompass both traces, allowing each trace to influence the other. As circuit elements, the magnetic coupling is represented by a mutual inductance L_{12}, while the electric coupling is represented by the mutual capacitance C_{12}.

The undesirable coupling of energy from a switching line (the *culprit*) to a passive line (the *victim*) is called *crosstalk* (also sometimes called *cross coupling*). This is what often first comes to mind when designers consider closely spaced traces running parallel to each other. A switching culprit line inducing noise voltage on a passive (unswitched) victim is discussed in Section 8.5.

A second case requiring analysis occurs when all of the signals in a closely spaced group (such as a wide data bus) switch simultaneously, with none being passive. The switching pattern of these signals results in a change of impedance (and, at least in microstrip, a change in the time of flight as well) of these signals.

Although this behavior is a result of coupling between traces (i.e., crosstalk), for emphasis it's given the term *simultaneous switching impedance variation* in this book and is discussed in Section 8.3. This in turn leads to a related topic, differential impedance, which along with differential signaling is discussed in Section 8.4.

8.2 Odd and Even Modes

Two transmission lines are shown linked by mutual inductance and capacitance in Figure 8.2. Both traces are driven by identical pulse generators (PG1 and PG2) having a fixed output impedance of R_g and are terminated in load resistors R_L such that reflections are not created when the launched waves reach the far end.

The pulse generators may either drive the lines in the same direction (i.e., in phase with one another) or in opposite directions (180° out of phase). The case where only one line switches while the other remains static is the crosstalk case discussed in Section 8.5.

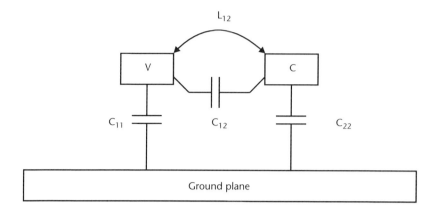

Figure 8.1 Coupling between side-by-side microstrips.

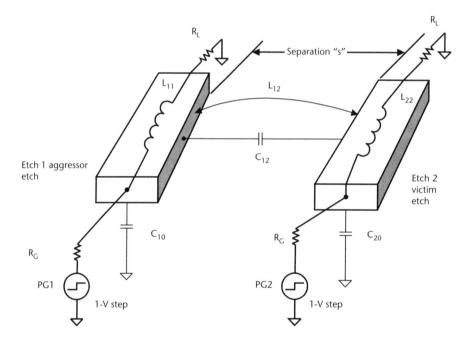

Figure 8.2 Two coupled parallel lines driven by separate sources.

Signals switching in the same direction and carrying the same current are called *even mode* signals because the electric field lines are symmetrically arraigned about an axis of symmetry [1], as shown for a microstrip in Figure 8.3. Stripline exhibits the same behavior but with the electric fields terminating on both plates.

When switching in opposite direction (*odd mode*), the electric field lines no longer exhibit this symmetry (although, as shown, the magnetic fields do).

8.2.1 Circuit Description of Odd and Even Modes

The behavior of coupling capacitance appearing between traces was examined in Chapter 3. The influence of mutual capacitance on a circuit was shown in

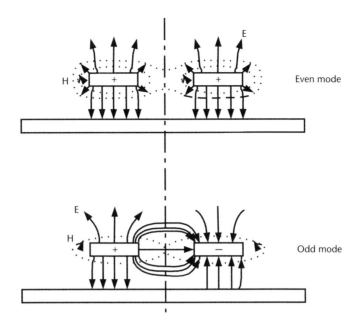

Figure 8.3   Even mode electric and magnetic fields.

Examples 3.3 and 3.4 to depend on the switching activity occurring between the traces. Traces carrying voltage switching in the same direction, with the same rate of change, effectively had zero mutual capacitance, as there was no charge transferred between the conductors. Alternatively, the mutual capacitance was seen to allow for charge transfer when the conductors switched in opposite directions.

Current flows in the same direction down the traces when switching in the even mode but flows in the opposite direction during odd-mode switching. Therefore in odd-mode switching, the loop inductance is reduced due to the advantageous application of mutual inductance (see Chapter 4), but it is not reduced in even-mode switching.

It's apparent the switching behavior of its neighbors will affect a trace's aggregate capacitance and inductance, and thus its impedance and time of flight. In fact, two separate impedances and times of flight exist in a two-conductor system: an odd-mode impedance (Z_{oo}), along with an odd-mode propagation delay (tpd_{odd}), and an even-mode impedance (Z_{oe}) and associated propagation delay (tpd_{even}).

Because in odd-mode switching, the trace's capacitance is maximized and the inductance is reduced by mutual inductance, the odd-mode impedance therefore is (8.1):

$$Z_{oo} = \sqrt{\frac{L_s - L_m}{C_s + C_m}} \qquad (8.1)$$

When propagating in the odd mode, waves have a delay per length as shown in (8.2):

$$tpd_{odd} = \sqrt{(L_s - L_m)(C_s + C_m)} \qquad (8.2)$$

In even-mode switching, mutual capacitance is essentially zero and the inductance is maximum.

Therefore, the even-mode impedance is (8.3):

$$Z_{oe} = \sqrt{\frac{L_s + L_m}{C_s - C_m}} \tag{8.3}$$

And the even-mode delay per length is (8.4):

$$tpd_{even} = \sqrt{(L_s + L_m)(C_s - C_m)} \tag{8.4}$$

As described in Section 3.5, the mutual capacitance C_m is usually reported as a negative value by field-solving software, but it's taken as positive in (8.1) through (8.4).

Equations (8.1) through (8.4) can be generalized for any number of conductors by making C_m and L_m the sum of all mutual capacitances and inductances. In the specific case of a two-conductor system, $C_s = C_{11}$ and $C_m = C_{12}$. Likewise for L_s and L_m.

Example 8.1

Using these L,C matrices, find Z_{oe} for:
(a) Conductor 1, assuming all conductors simultaneously switch high.
(b) Conductor 2, assuming all conductors simultaneously switch high.

The arraignment of conductors and a circuit schematic showing only the inductors appears in Figure 8.4. The self and mutual capacitances and resistance have been omitted from the schematic for clarity.

$$L = \begin{matrix} 10.5 & 3.11 & 1.05 \\ 3.11 & 10.3 & 3.11 \\ 1.05 & 3.11 & 10.5 \end{matrix} \quad \text{units of nH/in}$$

$$C = \begin{matrix} 3.38 & -1.00 & -0.042 \\ -1.00 & 3.73 & -1.00 \\ -0.042 & -1.00 & 3.38 \end{matrix} \quad \text{units of pF/in}$$

Solution

(a) The mutual terms from conductor 1 to conductor 2, found directly from the **L** and **C** matrices are simply summed:

$$\sum L_m = L_{12} + L_{13} = 3.11 + 1.05 = 4.16 \text{ nH}$$

$$\sum C_m = C_{12} + C_{13} = 1.0 + 0.042 = 1.042 \text{ pF}$$

Applying (8.3), $Z_{oe} = \sqrt{\dfrac{L_{11} + \sum L_m}{C_{11} - \sum C_m}} = \sqrt{\dfrac{10.5 \text{ nH} + 4.16 \text{ nH}}{3.38 \text{ pF} - 1.04 \text{ pF}}} = 79.15\Omega$

(b) Once again, the self and mutual terms are found directly from the matrices, noting that in this case they are relative to conductor 2.

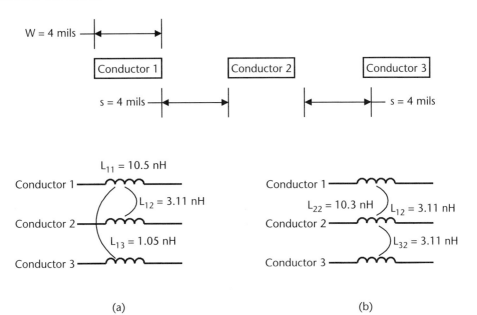

Figure 8.4   Circuit setup for Example 8.1.

$$\sum L_m = L_{21} + L_{23} = 3.11 + 3.11 = 6.22 \text{ nH}$$

$$\sum C_m = C_{21} + C_{23} = 1.0 + 1.0 = 2.0 \text{ pF}$$

$$\text{Again applying (8.3), } Z_{oe} = \sqrt{\frac{L_{22} + \sum L_m}{C_{22} - \sum C_m}} = \sqrt{\frac{10.3 \text{ nH} + 6.22 \text{ nH}}{3.73 \text{ pF} - 2 \text{ pF}}} = 97.7\Omega$$

It's clear that a trace's impedance is strongly influenced by the inductive and magnetic coupling between traces, and so for a multiconductor transmission line (such as the traces making up an address bus going from a microprocessor to a memory), the notion of a trace having one constant impedance is incorrect unless the spacing between traces is made large. In practical systems, the spacing between traces is usually made small to get the best possible signal routing density. As Example 8.1 demonstrates, each wire in such a bus will experience a variation in impedance as a function of the data pattern being transmitted. We'll return to this in Section 8.2.4.

8.2.2   Coupling Coefficient

The *coupling coefficient* (also called the *coupling factor*) provides a convenient way to gauge the degree of coupling between circuits. The magnetic coupling factor k_L was introduced in Chapter 4 and is repeated in (8.5):

$$k_L = \frac{L_{12}}{\sqrt{L_{11} L_{22}}} \tag{8.5}$$

where L_{12} is the mutual inductance between inductors L_1 and L_2, and L_{11} and L_{22} are the self inductance of the two conductors as described in Section 4.6.

In a similar fashion, the a capacitive coupling coefficient k_c is defined in (8.6):

$$k_C = \frac{C_{12}}{\sqrt{C_{11}C_{22}}} \qquad (8.6)$$

The coupling coefficient is a unitless number with an absolute value ≤ 1. A circuit is said to be *loosely coupled* when $|k|$ is "much less" than 1 ($|k| \ll 1$), and the coupling becomes increasingly *tight* as $|k|$ approaches 1. It can be shown [2] that the magnetic and capacitive coupling factors are equal for TEM lines such as striplines but will not be equal for microstrips.

Intuitively, closely spaced traces will be more tightly coupled than ones spaced further apart. This is apparent in Figure 8.5, which shows the coupling factor between a pair of 50-Ω and 65-Ω striplines for edge-to-edge spacing ranging from 2.5 mils to 25 mils. The traces are half ounce, 5 mil wide on FR4. As expected, the coupling coefficient k is seen to fall as the spacing between traces increases.

Regardless of the trace separation in this example, the 65-Ω trace has higher coupling than the 50-Ω trace, and in general for a given trace width, thickness, and ε_r, coupling will be higher for higher impedance trace, especially if the separation is not great. The trace dimensions in Figure 8.5 give a clue as to why this is so. The self capacitance must decrease to raise the impedance from 50Ω to 65Ω. As described in Chapter 3, to do so while holding w, t, and ε_r constant requires the spacing h to

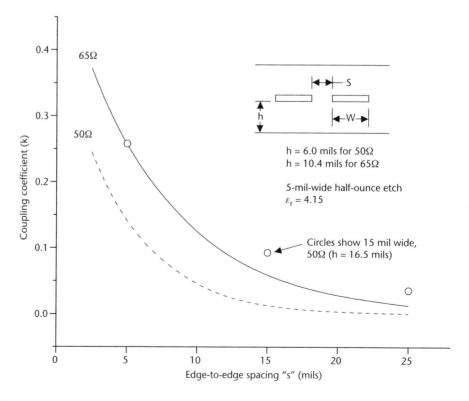

Figure 8.5   Coupling factor for half-ounce, 5-mil-wide stripline on FR4.

increase, which (assuming the same trace spacing) provides more opportunity for field lines to terminate on the adjacent trace rather than on the ground planes. As illustrated in the figure, for traces on a 5-mil spacing, C_{11} decreases by 20% when going from 50Ω to 65Ω while C_{12} increases by some 44%. From (8.6), the net result is an increase in K_c (and, incidentally, because in this example the trace is stripline, K_L will increase as well).

However, Figure 8.5 shows only part of the picture. Table 8.1 shows the coupling for various spacings with respect to the ratio of the trace's width for a 50-Ω 5-mil-wide and 15-mil-wide half-ounce stripline.

When viewed in this way, it becomes apparent that for a given space-to-width ratio, the coupling is higher for the 5-mil trace than for the 15-mil trace. In fact, the table points out an inconsistency with the commonly held rule (for example, see [3]) that traces should be spaced 2W (i.e., an edge-to-edge spacing two times its width, or $\dfrac{s}{w} = 2$) or greater to minimize coupling (and thus, as we'll see, crosstalk). Unfortunately, this same rule is also sometimes referred to as the *3W rule*, where the traces are measured center to center. As shown by the table, the efficacy of such a rule diminishes for narrow trace. For example, a pair of traces 5 mils wide separated by 10 mils edge to edge $\left(\dfrac{s}{w} = 2\right)$ has 1.5 times the coupling of a pair of 15-mil-wide lines separated by 30 mils.

It can be shown [4] that for loosely coupled lines (where the mutual capacitance and inductances are much less than the self capacitance and inductance, and so K_C and K_L are both << 1) $Z_{oo} < Z_o < Z_{oe}$, and that the impedance is (8.7):

$$Z_o \approx \sqrt{Z_{oo} Z_{oe}} \tag{8.7}$$

8.2.3 Stripline and Microstrip Odd- and Even-Mode Timing

Not apparent in (8.2) and (8.4) is a characteristic of TEM propagation: when propagating waves in a homogeneous dielectric (such as with stripline trace), the odd and even modes propagate at the same speed (i.e., $tpd_{odd} = tpd_{even}$). This follows from the discussion in Section 5.4.2, which showed that signals propagate with a velocity equal to the speed of light in the dielectric. It follows that in stripline, the two modes will propagate at the same speed because, regardless of the mode, all of the field lines are contained within the dielectric. However, this is not true with microstrip traces, where some of the field lines propagate in the laminate (dielectric)

Table 8.1 Coupling of a Narrow and Wide 50-Ω Stripline

$\dfrac{s}{w}$	$K_{5\,mil}$	$K_{15\,mil}$
1	0.124	0.093
2	0.035	0.023
3	0.010	0.006
4	0.003	0.001

and the remainder propagate in air (or, more typically, solder mask and then air). In that case, the $tpd_{odd} \neq tpd_{even}$ because the concentration of field lines in the air and dielectric changes depending on the mode [2, 5].

At the circuit-board level, this means the delay per unit length will depend on the switching activity of neighboring traces when propagating along microstrips but will be constant for stripline. However, it's shown in Section 8.3 how improper termination affects the timing of stripline circuits.

Example 8.2

Compute the Z_{oo}, Z_{oe}, Z_o and delay per inch and the coupling coefficients K_L and K_C for the two-conductor system represented by the **L** and **C** matrices shown in (8.8). The inductances are in nH/in; the capacitances are in pF/in.

$$\mathbf{L} = \begin{array}{cc} 8.55 & 0.302 \\ 0.302 & 8.55 \end{array} \tag{8.8a}$$

$$\mathbf{C} = \begin{array}{cc} 3.49 & -0.123 \\ -0.123 & 3.49 \end{array} \tag{8.8b}$$

Solution

From (8.1) and (8.2), the odd-mode impedance and delay is:

$$Z_{oo} = \sqrt{\frac{L_{11} - L_{12}}{C_{11} + C_{12}}} = \sqrt{\frac{8.55 \text{ nH} - 0.302 \text{ nH}}{3.49 \text{ pF} + 0.123 \text{ pF}}} = 47.8\Omega$$

and

$$tpd_{odd} = \sqrt{(L_{11} - L_{12})(C_{11} + C_{12})} = 173 \text{ ps/in}$$

And from (8.3) and (8.4), the even-mode impedance and delay is:

$$Z_{oe} = \sqrt{\frac{L_{11} + L_{12}}{C_{11} - C_{12}}} = \sqrt{\frac{8.55 \text{ nH} + 0.302 \text{ nH}}{3.49 \text{ pF} - 0.123 \text{ pF}}} = 51.3\Omega$$

and

$$tpd_{even} = \sqrt{(L_{11} + L_{12})(C_{11} - C_{12})} = 173 \text{ ps/in}$$

From (8.7), $Z_o \approx \sqrt{Z_{oo} Z_{oe}} = \sqrt{47.8 \times 51.3} = 49.5\Omega$

From (8.5) and (8.6) the coupling coefficients are:

$$k_L = \frac{L_{12}}{\sqrt{L_{11} L_{22}}} = \frac{0.302 \text{ nH}}{\sqrt{8.55 \text{ nH} \times 8.55 \text{ nH}}} = 0.035$$

and

$$k_C = \frac{C_{12}}{\sqrt{C_{11} C_{22}}} = \frac{0.123 \text{ pF}}{\sqrt{3.49 \text{ pF} \times 3.49 \text{ pF}}} = 0.035$$

Because the odd- and even-mode delays are the same, it's correct to conclude that (8.8) represent conductors within a homogeneous dielectric such as stripline. The coupling coefficients also verify this, as $K_C = K_L$ only when the dielectric is homogeneous, such as stripline.

Example 8.3

Recalculate Z_{oe}, Z_{oo}, tpd_{even}, tpd_{odd}, K_C, and K_L using the **L**, **C** matrices shown in (8.9). The inductances are in nH/in; the capacitances are in pF/in.

$$\mathbf{L} = \begin{matrix} 7.26 & 0.401 \\ 0.401 & 7.26 \end{matrix} \tag{8.9a}$$

$$\mathbf{C} = \begin{matrix} 2.98 & -0.041 \\ -0.041 & 2.98 \end{matrix} \tag{8.9b}$$

Solution

In the same manor as Example 8.2:

$Z_{oo} = 47.67\Omega$ and $tpd_{odd} = 144$ ps/in
$Z_{oe} = 51.07\Omega$ and $tpd_{even} = 150$ ps/in
$Z_o = 49.3\Omega$
$K_C = 0.014$ and $K_L = 0.055$

The different propagation times of the odd- and even-mode signals suggests (8.9) represents a system of propagation involving more than one dielectric. The capacitive and inductive coupling coefficients not being equal also support this conclusion. In fact, (8.9) represents a microstrip not covered by solder mask. Chapter 9 shows in more detail how solder mask affects microstrips' electrical characteristics. Although the total coupling is small in this example, the inductive coupling is seen to be nearly four times larger than the capacitive coupling. As is shown in Section 8.5, this has ramifications for the nature of the crosstalk voltages that a culprit will induce onto a victim.

8.2.4   Effects of Spacing on Impedance

Because it's the coupling between traces that is causing the odd- and even-mode impedances to differ from the nominal impedance, one would expect that increasing the spacing would cause the impedances to approach the value that a single trace would have in isolation. Figure 8.6 shows the odd- and even-mode impedances as a function of edge-to-edge spacing for a pair of 5-mil-wide, half-ounce 50-Ω and 65-Ω stripline (solid curve) and microstrip (dashed curve, computed without solder mask) traces on FR4.

It's evident that for small separations where the coupling is highest, there is a great difference between the odd- and even-mode impedances. As the separation

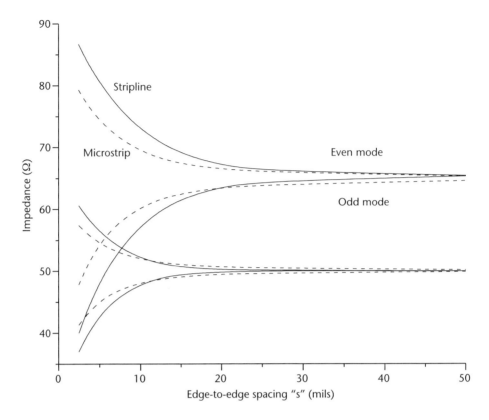

Figure 8.6   Odd- and even-mode impedance for a 5-mil-wide, half-ounce 50-Ω and 65-Ω (nominal) stripline (solid curve) and microstrip (dashed curve) on FR4.

increases, the even-mode impedance falls and the odd-mode impedance increases as they approach the nominal value.

Also apparent in Figure 8.6 is the relationship between the nominal impedance and the difference between the even- and odd-mode impedances. For any given separation, the 65-Ω trace is seen to have a larger impedance variation than the 50-Ω trace, even though both traces are half ounce, 5 mils wide on the same FR4 substrate. This is consistent with Figure 8.5, and as was described with the coupling coefficient, for a given ε_r and trace width, raising the impedance is accomplished by increasing the distance h. Doing so increases coupling between traces, making the variation between Z_{oo} and Z_{oe} larger than if the coupling were less.

8.3   Multiconductor Transmission Lines

It's known from elementary transmission line theory that to avoid reflections from the far end, a single wire transmission line should be terminated at the far end to a reference (generally ground or to a *reference voltage*, often called V_{tt}) in an impedance equal to Z_o. However, we've just seen that a system of two wires supports two propagation modes, each with their own impedance. In general, a system of N conductors has N modes of propagation [5], each with unique impedance. This suggests that a single resistor to ground at the far end of each line would not properly terminate a multiconductor line under all switching conditions. In fact, a network of resistors between each conductor and V_{tt} is necessary for ideal termination [6].

While quite acceptable as a way to terminate a pair of wires (such as a differential pair, covered in Section 8.4.2), applying this scheme to a wide bus consisting of many conductors is not feasible at the PWB level. Instead, the general practice for far-end termination is to use a single resistor with a value equal to Z_o connected V_{tt} (which could be ground). Although not ideal, a single resistor to can adequately terminate a multiconductor transmission line, especially if the spacing between traces is large. Large spacing means that the coupling coefficient k is small, which implies that $Z_o \approx Z_{oo} \approx Z_{oe}$, so a resistor having a value computed with (8.7) would be a good compromise under all switching conditions.

8.3.1   Bus Segmentation for Simulation Purposes

A SPICE simulation can be used to test the effectiveness of a termination scheme, but if the bus is very wide it's impractical (and unnecessary) to include all of the conductors in the simulation. Because k falls with increasing distance, it's reasonable to assume only a few neighboring conductors are necessary to model the reflection and termination behavior of the entire bus.

We'll use the coupling of a 65-Ω five-conductor stripline transmission line to demonstrate the magnitude of coupling between traces and thus to determine the number required in an example simulation. Sixty-five ohms is a commonly used PWB impedance when signaling at lower speeds (e.g., LVCMOS [7]), and, as was shown in Figure 8.5, in general k increases with higher impedance. Therefore, we would expect the coupling between neighbors to be higher than if a lower impedance (such as 50Ω) were chosen. High coupling is desirable in this example to better illustrate the intended concepts. The traces are shown in Figure 8.7 and are half-ounce copper, 5 mils wide on a laminate with $\varepsilon_r = 4.15$. The edge-to-edge spacing is shown as s.

The coupling coefficient from trace 1 to the other traces for values of $s = 5$ and $s = 25$ mils is shown in Table 8.2.

As expected, coupling falls off with distance, and by conductor 3 (the second conductor away), k has fallen below 0.1 for the case where $s = 5$ mils. It's negligible by the fourth conductor, and in the 15-mil case is essentially zero by the third conductor and beyond. This validates the intuitive assumption that for simulation and analysis purposes, it's only necessary to include some of the wires in a wide bus because the reach of a culprit trace is limited to its immediate neighbors. It also

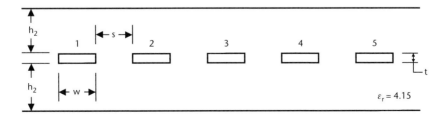

Figure 8.7   Five-conductor 65-Ω stripline.

Table 8.2 Coupling Coefficient
k from Trace 1 for $s = 5$ mils and
$s = 25$ mils

Conductor	k $(s = 5)$	k $(s = 25)$
2	0.236	0.012
3	0.064	2×10^{-4}
4	0.018	3×10^{-6}
5	0.005	4×10^{-8}

demonstrates that the number of conductors to be included depends on the spacing between them. In the $s = 5$ mils case, using the stackup shown in Figure 8.7, a system of seven conductors (three on either side of a center conductor) would be an excellent representation (in fact, a five-conductor system would suffice for most analysis), and three conductors (one on either side) would be quite sufficient in the 25-mil case.

8.3.2   Switching Behavior of a Wide Bus

A circuit schematic used to simulate the switching and termination behavior of a seven-conductor 65-Ω stripline appears in Figure 8.8. It consists of a center conductor flanked by three neighbors on each side. The spacing between traces is adjusted to be either $s = 5$ or $s = 25$ mils. All traces are half an ounce thick, 5 mils wide, and 4 in long, and they are driven from pulse generators switching 2.5V with a fixed 25-Ω output impedance. Each line is terminated in a 65-Ω resistor to a termination voltage V_{tt}, and a simple lumped RLC model is included to mimic the load represented by the micropackage and input circuits of a high-performance ASIC. This simple setup is adequate for this discussion but is not detailed enough to accurately predict timing or reflections.

The odd- and even-mode impedances of the center conductor can be calculated as described in Section 8.2.1. Doing so for the case where $s = 25$ mils yields $Z_{oo} = 64\Omega$ and $Z_{oe} = 66\Omega$. This is consistent with the data in Table 8.2, which shows k as being negligible in this case and thus suggests $Z_{oo} \sim Z_{oe}$. Clearly, with such low coupling, adjacent conductors have only a small effect on one another, so the switching activity of neighboring traces will not greatly affect the impedance. From (8.7), a resistor at the far end of each line to V_{tt} with a value equal to 65 should adequately terminate this transmission line for any data pattern present on neighboring traces.

However, k is not small for the $s = 5$ mils situation. In this case, $Z_{oo} = 40\Omega$ and $Z_{oe} = 105\Omega$. Once again using (8.7), a far-end resistor equal to 65Ω is the best compromise, but simulation is necessary to determine if that compromise is good enough for a specific application. It's worth noting that the 65-Ω variation in odd- and even-mode impedances when seven lines are included is quite a bit larger than the ~ 25-Ω variation of the two-conductor $s = 5$ mil stripline case shown in Figure 8.6. It's clear with the closer spacing that the switching activity of nonadjacent neighbors is significant, and to use only two signals in the simulation would yield misleadingly optimistic results.

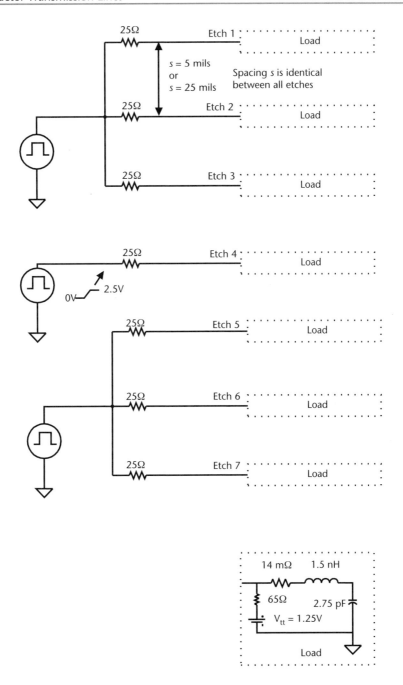

Figure 8.8 Seven-conductor stripline used in SPICE simulation.

8.3.3 Simulation Results for Loosely Coupled Lines

The near- and far-end simulation results of conductor number 4 for $s = 25$ mils appears in Figures 8.9 and 8.10. Waveforms are shown for even mode (all signals simultaneously switching high) and odd mode (signals on conductors 1–3 and 5–7 are driven low, while conductor 4 is driven high). As shown in Figure 8.8, all seven lines have a reactive load representing a micropackage and input receiver and are terminated in 65-Ω resistors to V_{tt}.

The near-end waveforms are a good place to observe the effects of switching activity on transmission-line impedance, as the plateau voltage (i.e., the voltage initially launched down the line) is determined by simple voltage divider action between the generator and transmission-line impedances. The generator impedance is fixed at 25Ω in this setup, making it easy to relate the change in transmission-line impedance directly with plateau voltage.

The difference between odd-mode and even-mode impedance is small with $s = 25$ mils, so the launched voltage should be nearly the same for any switching pattern. The voltage launched when switching in the even mode will be higher than when switching in the odd mode, as the even-mode transmission-line impedance is the higher of the two. The simulation results depicted in Figure 8.9 shows this to be the case: the even-mode launched voltage is slightly higher (15 mV) than the odd mode. This suggests the voltage received at the load should not change much due to the switching activity of neighboring traces, and the far-end simulation results presented in Figure 8.10 shows this to indeed be so. The absence of strong overshoots or ring back in the received waveform attest to the lines being properly terminated.

8.3.4   Simulation Results for Tightly Coupled Lines

As we saw, decreasing the separation s from 25 mils to 5 mils increases the coupling between traces, which results in a large variation between Z_{oo} and Z_{oe}. It's reasonable to assume such a line would be difficult to satisfactorily terminate at the far end with a single, fixed resistor to V_{tt} because the line's impedance (and hence, the proper value for the termination resistor) changes with the data being switched on neighboring traces. Another factor to consider is that voltage divider action between the

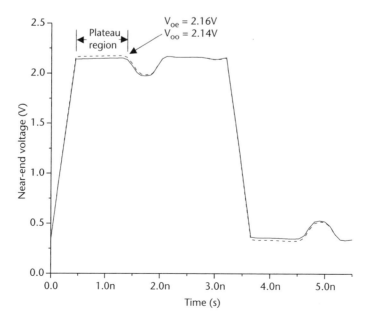

Figure 8.9   Near-end SPICE simulation results for trace number 4 in Figure 8.8 with $s = 25$ mils.

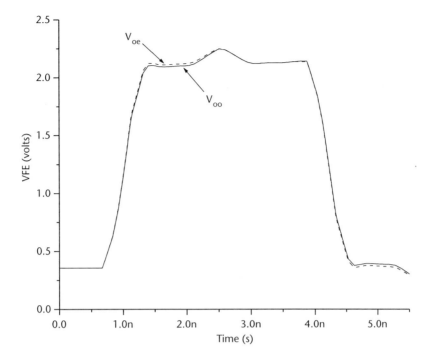

Figure 8.10   Far-end SPICE simulation results for trace number 4 in Figure 8.8 with *s* = 25 mils.

line's impedance and the generator's output impedance determines the voltage launched down the line. This means the launched voltage will also be data-pattern dependent, further complicating the load waveform.

The launched voltage data dependence is evident in Figure 8.11, which shows the plateau voltage at the near end of conductor 4 using the setup depicted in Figure 8.8 with *s* = 5 mils. Conductor 4 represents the worst case, as it's aggressed on both sides by three conductors (for a total of six aggressors), whereas a conductor located closer to the edge (such as conductor 1) is not significantly affected by conductors more than four away (see Table 8.2).

The wide impedance range results in a large difference between the odd- and even-mode launched voltages, suggesting that in this case there will be a greater variation in the far-end voltage than shown in Figure 8.10 for the *s* = 25 mils case. This is confirmed in Figure 8.12.

The far end voltage characteristics appearing on trace number 4 are seen to depend strongly on the switching activity of neighboring traces. The included graphic shows traces 1–3 and 5–7 each as one conductor for simplicity. The far-end signal is well behaved in the nominal case where trace 4 switches high while all the other traces remain unswitched. Overshoot is evident in the odd mode, causing the voltage to peak at a higher value than in the even-mode switching case, even though from Figure 8.11 the launched voltage is the lowest in the odd-mode case. This peaking is caused by the large positive reflection voltage created by the impedance mismatch appearing between the 65Ω-termination resistor and the 40Ω-transmission line impedance. In the even-mode waveform, the flat portion is due to the negative reflection voltage that is created by the termination impedance being so much lower than the 105-Ω transmission line impedance.

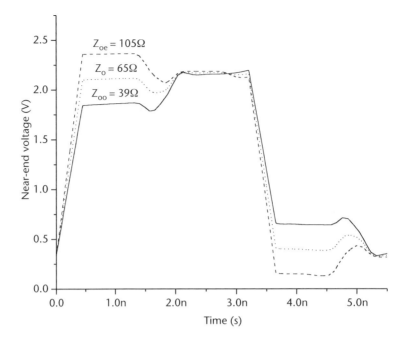

Figure 8.11   Near-end SPICE simulation results for trace number 4 in Figure 8.8 with $s = 5$ mils.

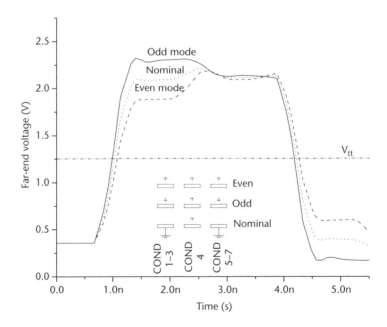

Figure 8.12   Far-end SPICE simulation results for trace number 4 in Figure 8.8 with $s = 5$ mils.

8.3.5   Data-Dependent Timing Jitter in Multiconductor Transmission Lines

In Section 8.2.1 it was shown that for stripline $tpd_{odd} = tpd_{even}$, but that in microstrip the two modes had different delays per unit length. This suggests that for microstrip, the time when a received signal crosses a given reference voltage would depend on the mode, but for stripline the received signal would always cross the reference voltage at the same instant, regardless of mode. While this assumption holds true for microstrip, it does not always hold true for stripline.

It's apparent from the stripline response shown in Figure 8.12 that the received signal crosses V_{tt} at different times, depending on the mode. The *nominal* case is the response when only the center conductor switches and all other conductors are held low. In this example, the even-mode signal crosses V_{tt} later than the nominal case, and the odd-mode signal crosses sooner. The odd-to-even-mode timing difference in this example is an 80-ps data-dependent uncertainty (*jitter*) in the waveform's position that would need to be accounted for when analyzing a systems setup or hold timing margins. This jitter does not come about because of a change in the transmission lines delay but instead is an artifact of the chosen termination. As demonstrated in Example 8.2, the odd- and even-mode delays are the same for stripline. This is evident in Figure 8.12, which shows the odd- and even-mode signals starting to rise at the same instant. However, a single resistor can only terminate the nominal switching case and as shown allows reflections to be created in the odd- and even-mode cases. The reflections combine with the incident waveform being received and cause the rate at which the received signal rises to be different for each mode. In Figure 8.12, the received odd-mode signal has a faster slew rate and thus crosses the reference voltage sooner than the even-mode signal. This phenomenon is exacerbated in microstrip and can be avoided by proper termination of both modes.

8.4   Differential Signaling, Termination, and Layout Rules

Differential signaling—using two wires to simultaneously send the true and complement versions of a signal—is a commonly used high-speed interconnect technique. Although it requires twice as many wires, when properly implemented, differential signaling has better noise immunity than single-ended signaling (i.e., signals referenced to a voltage assumed to be common between the transmitter and receiver, such as ground or a reference voltage). In fact, a characteristic of differential signaling is that it allows proper reception even when the transmitter and receiver grounds are not at the same voltage. This has great advantage when signaling across long cables or between cards plugged into a backplane because in both cases there will be a voltage difference between the transmitter and receiver local grounds. And, in both cases crosstalk between signals or noise coupled in from the return path will alter the switching levels, degrading performance. To a certain extent, differential signaling is immune to both of these effects, as discussed next.

8.4.1   Differential Signals and Noise Rejection

Differential signaling uses a differential transmitter and differential receiver arraigned as shown in Figure 8.13.

The transmitter simultaneously sends out voltages corresponding to a true and complement version of the input signal on the V_p and V_n wires, respectively. The signal is seen to swing above and below a common mode voltage V_{cm} and has a

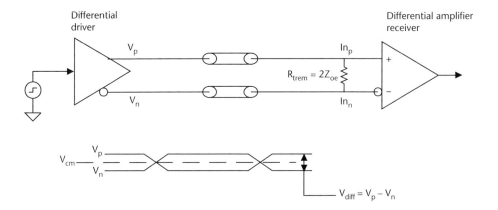

Figure 8.13   Differential transmitter and receiver.

differential voltage V_{diff} that is the difference between the voltage on the positive input and the negative input as shown in (8.10):

$$V_{diff} = V_p - V_n \qquad\qquad (8.10)$$

The receiver is a differential amplifier that rejects voltages common to both inputs (*common-mode noise*) but amplifies the voltage difference between the inputs. The amplifier has sufficient differential gain so that when a specified differential voltage is present on the inputs, the output switches rail to rail (i.e., provides a valid logic "1" or "0" on its output).

Differential signaling is unique in that the receiver responds to the voltage difference between two signals, not to the value of the voltage with respect to the local reference plane (generally ground). Noise voltages simultaneously coupled onto In_p and In_n become common to both (common-mode noise) and (within specified limits) will be rejected by the receiver. This behavior is fundamental and provided the signals are within the proper *common-mode range*, the receiver will properly distinguish between valid logic states if the differential voltage is adequate. From a receiver design perspective, the common-mode range specification is required to allow node voltages within the differential amplifier receiver to stay properly biased. Input signals resulting in compromised internal biasing causes the receiver's timing to become affected, reduces its ability to reject common-mode noise, and, if large enough, causes the receiver to improperly distinguish between logic states.

8.4.2   Differential Impedance and Termination

Differential amplifier input terminals have high input impedance, requiring termination at the receiver inputs to impedance match it to the line. The value should be equal to the *differential impedance* of the line so that the proper differential voltage is developed across it without creating reflections. In practice, the termination is either placed internal to the receiver itself or externally on the PWB near the receiver input terminals.

The intuitive approach shown in Figure 8.14 will first be used as a precursor to a more rigorous analysis of differential impedance and termination.

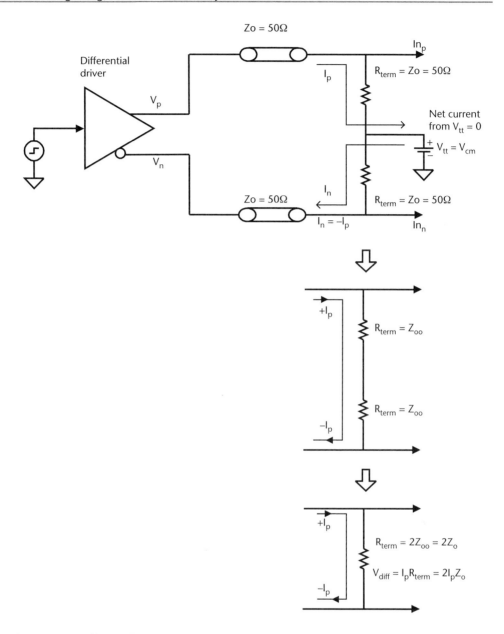

Figure 8.14   Differential termination.

As shown, the differential driver transmits out a voltage V_p on the positive output and V_n on the negative. These voltages cause currents I_p and I_n to flow down the transmission lines forming the differential pair. By definition, a differential pair operates in the odd mode because (ideally) the two signals are always 180° out of phase. Therefore, for proper termination, each line must be terminated by a resistor R_{term} equal to Z_{oo} to V_{tt}, where V_{tt} is the common-mode voltage appearing between V_p and V_n. Assuming edge-coupled traces, if the trace's edge-to-edge spacing is very large $Z_{oo} \approx Z_o$, and (as is often done in practice) R_{term} equals Z_o.

Because the common-mode voltage is centered between V_p and V_n, the currents I_p and I_n are identical. They flow in opposite directions, making the net current flow from the V_{tt} supply zero. Because the current flow is zero, the supply can be removed without disturbing the circuit's operation. The two termination resistors are seen to be in series and so can be replaced with a single resistor equal in value to twice Z_{oo}.

With this intuition in mind, we now turn to a more formal analysis.

From network theory (for example, see [8]), a *mutual impedance* is said to exist between conductors if a current in one conductor is able to induce a voltage in the other. Employing array notation as used in Chapters 2 through 4 to describe the resistance, capacitance, and inductance matrices, Z_{11} is the self impedance of conductor 1 when all other conductors are open circuited. The mutual impedance between conductors 1 and 2 is shown as Z_{12}.

The voltages V_p and V_n in Figure 8.14 are found by multiplying the self and mutual impedances by I_p and I_n, as shown in (8.11):

$$V_p = Z_{11} I_p + Z_{12} I_n \tag{8.11a}$$

$$V_n = Z_{22} I_n + Z_{21} I_p \tag{8.11b}$$

From (8.10), the differential voltage is the voltage difference between V_p and V_n. Therefore, the differential voltage is (8.12):

$$V_{diff} = Z_{11} I_p + Z_{12} I_n - Z_{22} I_n - Z_{12} I_p \tag{8.12}$$

Setting $I_n = -I_p$ and $Z_{21} = Z_{12}$, recombining and simplifying (8.12) yields (8.13):

$$V_{diff} = 2I_p (Z_{11} - Z_{12}) \tag{8.13}$$

The differential impedance Z_{diff} is the ratio of the differential voltage to the current flowing through the line, as shown in (8.14):

$$Z_{diff} \equiv \frac{V_{diff}}{I_p} = \frac{2I_p(Z_{11} - Z_{12})}{I_p} = 2(Z_{11} - Z_{12}) \tag{8.14}$$

The odd-mode impedance Z_{oo} and even-mode impedance Z_{oe} are readily found from the self and mutual impedances as shown in (8.15) and (8.16).

$$Z_{oo} = Z_{11} - Z_{12} \tag{8.15}$$

$$Z_{oe} = Z_{11} + Z_{12} \tag{8.16}$$

Because from (8.15)

$$Z_{oo} = (Z_{11} - Z_{12}), \ Z_{diff} = 2Z_{oo} \tag{8.17}$$

The results in (8.17) match the intuitive analysis presented in Figure 8.14 and show that to properly impedance match a differential line, the value of R_{term} in Figure 8.13 should be twice the odd-mode impedance of a pair of coupled lines. Notice that if the coupling is small enough, $Z_{oo} = Z_o$, and so $R_{term} = 2Z_o$, which is the value often used in practice.

The analysis to this point has assumed the differential signals are truly odd mode, with no even-mode component. This is rarely the case: signals will often have at least a small common-mode component that causes the signal to be less than perfectly differential. Such errors arise from differential drivers that are not perfectly matched in slew rate or amplitude so that the true and compliment signals are not exact mirror images of one another. Another source of error is in routing the signal pairs so they are not absolutely identical. If this happens, noise that is not symmetrically coupled onto both lines will cause current to flow in one line but not the other, such that $I_p \neq I_n$, invalidating the previous analysis. It's especially easy to inadvertently violate this rule when routing the signal through connectors or other dense pin fields by forming the diff pair using edge-coupled traces. An example of this appears in Figure 8.15.

As shown, the differential pair $A1$ and $-A1$ is brought out of the connector pin field by paths that are quite different. The signal trace connected to the $A1$ pin passes between pins B through D carrying true and complement signals. If the signals on these pins are truly differential, as they switch the net charge transferred from them to the $A1$ trace is zero. This is not the case with the complement: the $-A1$ trace has static ground pins on one side and pins switching on its other side. Therefore, unlike the $A1$ signal, the $-A1$ signal will not be symmetrically aggressed when pins B through D switch. This difference makes the $A/-A1$ signal pair not truly differential and the signal would suffer from data-dependent jitter.

In most practical cases, a single resistor from V_p to V_n is adequate. However, in those cases where the even mode signal is large enough, the termination scheme shown in Figure 8.16 (sometimes called a *Pi termination*) that properly terminates both modes [6] must be used.

In the Pi terminator, resistors are placed between each conductor and the reference voltage (which could be ground). Resistors R_1 and R_2 are computed with (8.18) [6]:

$$R_1 = Z_{oe} \tag{8.18a}$$

$$R_2 = \frac{2Z_{oe}Z_{oo}}{Z_{oe} - Z_{oo}} \tag{8.18b}$$

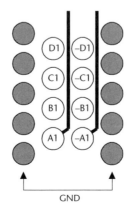

Figure 8.15 Wiring through dense pin field.

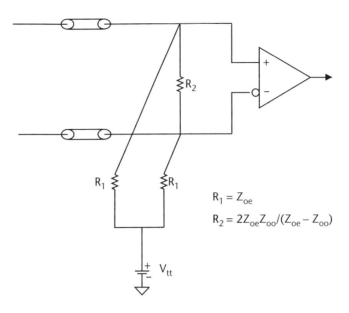

$$R_1 = Z_{oe}$$
$$R_2 = 2Z_{oe}Z_{oo}/(Z_{oe} - Z_{oo})$$

Figure 8.16   Pi termination network for a differential line.

Notice that if $Z_{oo} = Z_{oe}$ (which implies that the coupling coefficient k is very small, so $Z_o = Z_{oo} = Z_{oe}$), then R_2 becomes infinite, and the line is simply terminated in R_1 to V_{tt} as shown in the top portion of Figure 8.14 (or, equivalently, from In_p to In_n in $2R_1$, as shown in the bottom portion).

The scheme in Figure 8.16 is also useful when it's necessary to simultaneously rebias and terminate a signal at a receiver. An example of this situation is shown in Figure 8.17 and arises when the transmitter is of one logic type (3.3V CMOS, in this example, as might come from an oscillator), but the receiver is of a different technology (such as the receiver within an LVPECL [9] fan-out device). The series capacitors C_1 prevent a dc path from being present between the transmitter and receiver, thereby allowing each to operate at their own common-mode voltage. Resistors R_1 provide a path to V_{bias}, which sets proper common-mode voltage for the receiver. Resistors R_1 and R_2 are calculated with (8.18), again noting that if the lines are very loosely coupled, R_2 becomes infinite, leaving only R_1.

8.4.3   Reflection Coefficient and Return Loss

Energy is reflected from the load when the transmission line and load impedances are not identical (i.e., do not *match*). The degree to which a line and load are mismatched can be specified in several ways. Two commonly used metrics are the *voltage reflection coefficient* (in this book identified with ρ, to distinguish it from the complex reflection coefficient Γ), and *return loss* (RL). The reflection coefficient is convenient because it's applicable to lines transmitting signals such as pulses and is easily calculated from oscilloscope measurements or simulation of the incident and reflected voltage waveforms. The return loss shows the ratio of incident to reflected power at the load at a specific frequency and so indirectly indicates the load-to-line

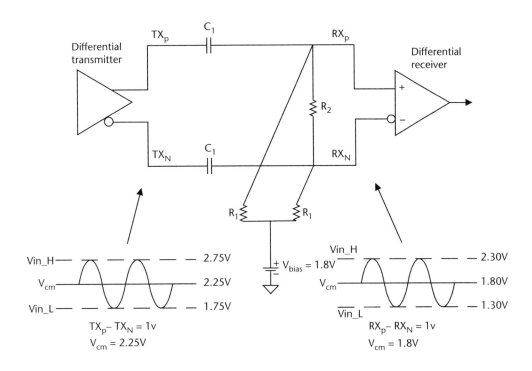

Figure 8.17   Using Pi network to rebias and terminate a receiver load.

mismatch. This way of specifying the load mismatch is popular in the high-speed serial interconnect industry.

The relationship between ρ and RL is presented in (8.19) and (8.20). The proofs are straightforward and are commonly available. For example, see [10].

The load voltage reflection coefficient ρ_L is given in (8.19) as a ratio of the load impedance (Z_L) and the transmission-line impendence (Z_o):

$$\rho_L \equiv \frac{V_{reflected}}{V_{incident}} = \frac{Z_L - Z_o}{Z_L + Z_o} \tag{8.19}$$

A voltage reflection coefficient of zero represents a perfectly matched line (i.e., $Z_o = Z_L$). A short circuit results in $\rho_L = -1$; an open circuit produces $\rho_L = +1$.

The return loss for a specific frequency is related to ρ_L as shown in (8.20) and is expressed in decibels.

$$RL = 20 \log \left(\frac{1}{|\rho_L|} \right) \tag{8.20}$$

A perfectly matched line has an infinite return loss because no power is reflected from the load. In that case $\rho_L = 0$ and $\log \left(\frac{1}{\rho_L} \right)$ becomes infinite.

Example 8.4

A high-speed SERDES chip designed for 100-Ω differential impedance interconnect has an actual differential input impedance of 80 at a certain frequency. Calculate ρ_L and RL.

Solution

From (8.19) $\rho_L = \dfrac{Z_L - Z_o}{Z_L + Z_o} = \dfrac{80 - 100}{80 + 100} = -0.111$.

From (8.20) $RL = 20\log\left(\dfrac{1}{|\rho_L|}\right) = 20\log\left(\dfrac{1}{0.111}\right) = 19.1$ dB.

8.4.4 PWB Layout Rules When Routing Differential Pairs

Referring to Figure 8.18, use the following guidelines when routing edge-coupled differential pairs:

- Maintain the proper differential impedance:
  - Keep the spacing *s* constant between D and –D.
  - Minimize layer changes (*layer hopping*)—if possible route the edge-coupled diff pair on a single layer.

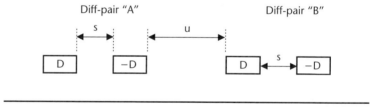

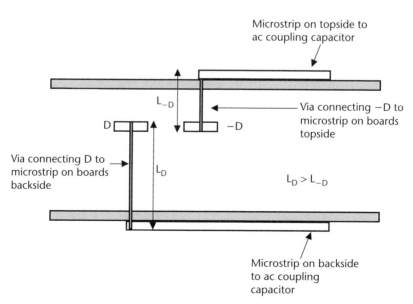

Figure 8.18   Signal trace length mismatch caused by unequal via lengths.

- Do not allow any other wires to come between the two wires forming the differential pair.
- Make the electrical environment identical for the wires forming the differential pair:
  - Route the diff pair on the same layer—do not route D on one layer and –D on different layers (by their very nature, this is not possible when routing broadside coupled pairs, as discussed in Chapter 9).
  - Ensure any external coupling that is present is applied equally to D and –D. Keep the noise-common mode; do not allow it to affect one line more than the other. See Figure 8.15 for an example of poor routing.
- Keep unrelated signals away from the diff pair:
  - Make the spacing $u \geq s$.
- Avoid return path splits (see Chapter 6):
  - Ensure the diff-pair return paths are proper.
  - If a power plane forms one of the return paths, ensure it's the same voltage powering the differential transmitter and receiver.
- Match the lengths of D and –D:
  - Each 5 mils of length difference between D and –D represents ~ 1-pS skew on FR4 stripline.
  - Vias used to bring the diff pair from a stripline layer to the surface must be the same length (e.g., placing capacitors to ac couple a diff pair on alternate sides of the board will cause one of the signals to be longer than the other).
  - Don't just match the total trace length—stripline and microstrip signals travel at different speeds. Therefore, to reduce skew, D and –D should have the same lengths of stripline and microstrip.
- Avoid vias:
  - Use as few vias as possible, as each represents a loss and an impedance discontinuity.
  - Place the vias transitioning D and –D adjacent to one another to encourage differential coupling. Be sure to maintain the proper differential impedance.
  - The D and –D lengths should be the same when a via is encountered.

8.5 Crosstalk

If strong enough, crosstalk—the undesired coupling of energy from one or more culprit lines to one or more victim lines—causes receiver noise margins to be reduced, thereby rendering a circuit susceptible to false triggering. Crosstalk-related problems can be particularly vexing to debug, especially if the circuits only fail occasionally.

A typical situation is shown in Figure 8.19. A clock sent between ASICs is routed adjacent to a data line with a spacing $s = 4$ mils for a distance of 12 in. The routing layer is 1-oz copper and is specified by the PWB vendor to have a 63-Ω

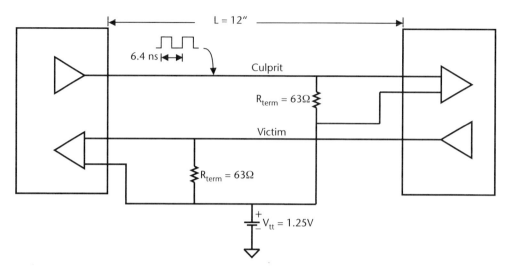

Figure 8.19 Aggressor/victim arraignment.

impedance (i.e., a line taken in isolation will have $Z_o = 63\Omega$) when the trace is 4 mils wide.

The victim data line is asserted low by its driver (which is series terminated and has an output impedance of 63Ω), and likewise is terminated to V_{tt} at its far end in 63Ω.

Coupling from the culprit to victim line causes noise pulses (crosstalk) to appear at both ends of the victim line. It's evident from the simulation results presented in Figure 8.20 that the coupling is strong and furthermore that the forward

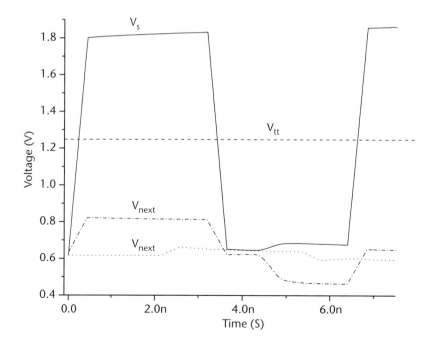

Figure 8.20 Simulation results of the setup in Figure 8.19.

crosstalk pulse—in this text called far-end crosstalk (FEXT)—has different charac-teristics than the crosstalk pulse traveling back to the load—near-end crosstalk (NEXT).

In this example, the NEXT and FEXT pulses are seen to have the same polarity as the aggressor voltage (V_s), but V_{next} has higher amplitude. The near-end noise pulse occurs coincident with the rising edge of the clock (V_s) and reaches its peak at the same time as V_s. The far-end pulse occurs later in time and has a more rounded rising edge than V_s. The circuit model presented next will help to explain this behav-ior. Simple formulas for calculating NEXT and FEXT voltages on lossless lines fol-low from that discussion.

8.5.1   Coupled-Line Circuit Model

A circuit model of the stripline used in Figure 8.19 appears in Figure 8.21. The line is divided into many infinitesimal segments dx in length [11], causing the R, L, and C terms to be scaled as shown.

Just the coupling portion of one of the segments depicted in Figure 8.21 is shown in Figure 8.22. To simplify, the notation $\Delta x C_{12}$ is replaced with C_m, and $\Delta x L_{12}$ is replaced with L_m. For this analysis, the line is assumed to have low losses so that R and G can be ignored.

A voltage launched down the culprit will cause a current i and a voltage v_s to travel out from the generator. As the voltage wave passes by, the mutual capacitance C_m will introduce a current i_c in the victim line proportional to the rate of change of the voltage (8.21):

$$i_c = C_m \frac{dv_s}{dt} \tag{8.21}$$

The mutual inductance L_m will cause a voltage v to be created on the victim line proportional to the rate of change of the current (8.22):

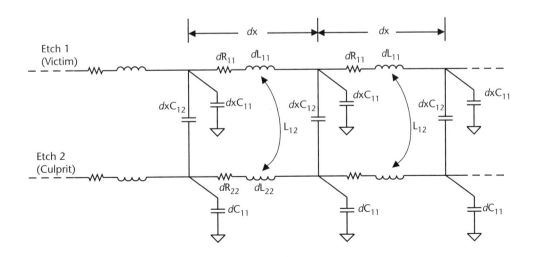

Figure 8.21   Coupled-line circuit model.

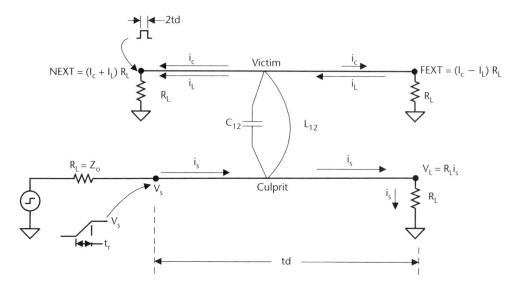

Figure 8.22 Mutual capacitance and inductance of one infinitesimal segment *dx*.

$$v = L_m \frac{di_s}{dt} \tag{8.22}$$

The induced voltage causes a current i_l to be introduced to the victim line in proportion to its impedance (8.23):

$$i_l = \frac{v}{Z_o} \tag{8.23}$$

The voltage at V_s causes a current i_s to travel down the culprit line toward the far end. As it does so, the mutual inductance causes a current i_l to flow in the victim line. From Lenz's law, this current will flow in the opposite direction, back toward the source end. This is in contrast to the current coupled to the victim by C_m. That current sees the same impedance on either side of the coupling point and so divides evenly, sending half back towards the source and half forward toward the load end. At the near and far ends, the two currents add and flow through the load resistors R_L, producing noise voltages.

Because the two currents are flowing in the same direction at the near end, when they add the NEXT pulse, it is seen to have the same polarity as V_s. However, the induced currents on the victim line are moving away from the direction that the signal is moving on the culprit line. This means that as the signal moves down the culprit, the current pulse it induces in the victim attaches itself to the end of the current pulse already making its way toward the culprit's near end. Said another way, the currents induced further up the line can't add to the previously induced ones because the earlier ones have moved back toward the near end by the time the currents further up the line are induced [11]. Provided the line is longer than the culprit voltage rise time (t_r), the result is a pulse of constant amplitude having a width equal to twice the line delay (*td* in Figure 8.22). The pulse will be truncated with shorter lines.

On the other hand, the FEXT pulse polarity is determined by the relative magnitudes of i_L and i_C. Because at the far end they are flowing in opposite directions, the two currents will cancel if i_L and i_C are equal (as occurs in lossless in stripline), producing no FEXT pulse.

In the forward direction, the currents will incrementally add along the length of the line, making the magnitude of FEXT grow in proportion to the length of the coupled line. The width of the FEXT pulse is equal to the rise time of the culprit signal, t_r.

The polarity of the FEXT pulse depends on whether the magnetic or capacitive coupling dominates. If $i_L > i_C$ the FEXT pulse will be opposite in polarity to the culprit signal, but (assuming the lines are properly terminated) the NEXT pulse will always be the same polarity as the culprit. The characteristics also depend on whether the pulse is traveling along microstrip or stripline trace. Because in stripline, the dielectric is homogeneous, the propagation is TEM (provided the losses are low) and the inductive and capacitive coupling coefficients are equal. This makes the currents induced by the magnetic and electric coupling cancel at the far end. For this reason, FEXT is zero for a properly terminated low-loss stripline but not zero for microstrip (regardless of loss).

Table 8.3 summarizes the NEXT and FEXT pulse characteristics, and Figure 8.22 schematically shows the induced current flow caused by the magnetic and electric coupling.

8.5.2   NEXT and FEXT Coupling Factors

It's possible to develop simple *crosstalk coupling factors* for NEXT and FEXT, provided the lines are lossless, loosely coupled, and properly terminated (so that reflections are not created). Loose coupling means the presence of the culprit line does not change the impedance of the victim line. This makes it straightforward to terminate these lines with a resistor equal to Z_o. Simple crosstalk formulas created under those assumptions are abundant in the literature (see, for example, [11–16]) and appear in (8.24) for K_b (the NEXT coupling factor) and (8.25) for K_f (the FEXT coupling factor):

$$K_b = \frac{1}{4\sqrt{L_{11}C_{11}}}\left(\frac{L_{12}}{Z_o} + Z_o\left|C_{12}\right|\right) \qquad (8.24)$$

Table 8.3   NEXT and FEXT Summary

	NEXT	FEXT
Amplitude	Independent of line length if $t_r < 2td$ (e.g., *long line*); appears as a miniature version of V_s; Increases with larger V_s	Increases with line length (up to $\frac{\lambda}{4}$) and increases with faster t_r and larger V_s
Width	Equal to $2td$ if $t_r < 2td$	Equal to t_r
Polarity	Always the same as V_s (provided no reflections)	Opposite polarity of V_s if $(L_{12} > C_{12}Z_o^2)$; will be zero if $K_c = K_f$ (i.e., stripline)

$$K_f = \frac{-1}{2}\left(\frac{L_{12}}{Z_o} - Z_o|C_{12}|\right) \tag{8.25}$$

where Z_o is defined as (8.26):

$$Z_o = \sqrt{\frac{L_{11}}{C_{11}}} \tag{8.26}$$

K_f has units of time per unit length (e.g., nanosecond per inch). It can be shown that K_b is dimensionless [12].

The loose coupling, low loss, and perfect termination assumed in the development of (8.24) and (8.25) are not usually found in practice, which limits their usefulness. Developing simple formulas to predict crosstalk in the presence of loss is especially difficult [17, 18]. Even in the simple lossless case, using (8.24) and (8.25) to predict crosstalk in systems of more than just a few conductors is usually not worth the manual effort. This is especially so if the actual lines are lossy, as skin effect losses can significantly change the FEXT response. Computer simulation using multiconductor transmission line models that incorporate skin-effect losses is far quicker and more accurate—and, besides accounting for loss, such models will properly predict reflections caused by imperfect termination or reactive loads (such as the parasitic capacitance of integrated circuit I/O pads). In fact, the principal modern use of (8.24) and (8.25) is to illustrate crosstalk's fundamental characteristics.

8.5.3   Using K_b to Predict NEXT

If the line is long (as is generally the case in high-speed PWB designs such that $t_r < 2td$), K_b can predict the NEXT voltage as shown in (8.27):

$$NEXT = K_b dV_s \tag{8.27}$$

where V_s is the voltage induced at the near end (see Figure 8.22). The width of the induced signal will be equal to a delay of twice the length of the line ($2td$), and for low loss or lossless lines will have the same rise time V_s. These calculations show very good to excellent agreement with SPICE simulation results and actual hardware measurement, even when the line is moderately lossy. Provided the line is long, from (8.27), NEXT amplitude is seen to depend on the aggressor's voltage but not its slope. The amplitude does not depend on the line's length but rather it *saturates* at a value given by (8.27).

However, if the line is short ($t_r \geq 2td$) the coupled voltage is also proportional to edge rate of V_s and is reduced over long line case by a factor of $\dfrac{2td}{t_r}$ as shown in (8.28):

$$NEXT = 2K_b td \frac{dV_s}{dt} \tag{8.28}$$

8.5.4   Using K_f to Predict FEXT

The far end crosstalk can be estimated with (8.29):

$$FEXT = K_f \, length \, \frac{dV_s}{dt} \qquad (8.29)$$

where $\frac{dV_s}{dt}$ is the slope of the culprit signal. The FEXT amplitude is seen to grow with line length and will be greater for high-voltage, fast-rise time signals.

Predicting FEXT with (8.29) is not nearly as successful as predicting NEXT with (8.28), especially if the lines are lossy. Additionally, any inaccuracy in predicting (or measuring, in a laboratory situation) the slope of the aggressor waveform $\left(\frac{dV_s}{dt} \right)$ will adversely affect the accuracy of the computed results. It's especially hard to predict the aggressor's slope without computer modeling, as its slope is usually dependent on the load impedance.

8.5.5   Guard Traces

Guard traces—placing a ground trace between conductors—can be effective in reducing crosstalk. Figure 8.23 shows one example of the effectiveness of guard traces.

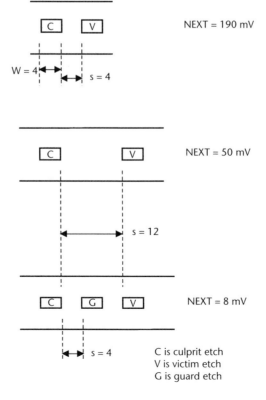

Figure 8.23   Example of stripline guard trace effectiveness.

The configuration appearing in Figure 8.19 viz., a pair of 4-mil-wide striplines separated by 4 mils, appears at the top of Figure 8.23. The NEXT voltage is 190 mV. In the middle portion of the figure, the NEXT voltage has been reduced to 50 mV by increasing the separation to 12 mils. Keeping the same pitch but adding a guard trace further reduces the noise to 8 mV.

On long lines it's important to ensure the guard trace is connected to ground at frequent intervals (such as every tenth of a rise time). If this is not done, the guard trace can become electrically long (i.e., becomes a transmission line). In this situation, energy coupled from the culprit into the guard trace will travel along the guard trace as it would any other transmission line, eventually reflecting off the ground connections appearing at its near and far ends. This energy is available to couple into the trace that the guard trace is intended to protect, degrading its shielding effectiveness.

A secondary effect of a well-connected guard trace is that the shielding makes the impedance of the transmitting trace less prone to impedance variations when neighboring signals switch—the difference between Z_{oo} and Z_{oe} becomes small. A *coplanar transmission line* (for example, see [19, 20]) is formed if guard traces are placed on both sides of the signals such that they surround the signal trace. Such a transmission line is very immune to coupled noise, especially if it's formed as stripline.

8.5.6 Crosstalk Worked Example

Example 8.5

The two conductor system appearing in Figure 8.19 is represented by the L,C matrices in (8.30a) and (8.30b). The matrices have units per inch. Compute the near- and far-end crosstalk if the lines are 12 in long and compare with the simulation results shown in Figure 8.20.

$$L = \begin{matrix} 10.515 \text{ nH} & 3.169 \text{ nH} \\ 3.169 \text{ nH} & 10.514 \text{ nH} \end{matrix} \tag{8.30a}$$

$$C = \begin{matrix} 3.375 \text{ pF} & -1.017 \text{ pF} \\ -1.017 \text{ pF} & 3.375 \text{ pF} \end{matrix} \tag{8.30b}$$

Solution

The circuit schematic of the setup appears in Figure 8.24. A driver launches a signal that switches between 0.62V and 1.8V in 450 pS down the culprit line. As shown, both ends of the victim line are terminated in 63-Ω resistors.

The first step is to find the crosstalk coupling coefficients K_b and K_f. Both require calculation of Z_o. From (8.26), $Z_o = 55.8\Omega$.

Applying (8.24) and (8.25), $K_b = 0.158$ and $K_f = 0$ ns/in.

Since $td > 2t_r$, for crosstalk purposes the line is long, and NEXT is calculated from (8.27) rather than (8.28) as:

$$V_{next} = K_b dv = 0.151(1.80V - 0.62V) = 0.178V$$

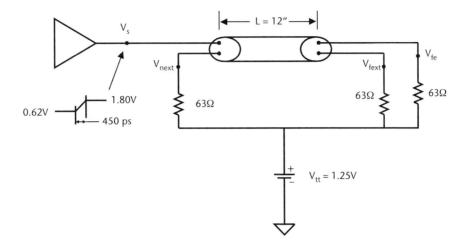

Figure 8.24 Circuit schematic of the two-coupled lines for Example 8.5.

Although the line was designed to be 63Ω, from (8.26) its impedance is in fact 55.8Ω. The 63Ω at the far end misterminates the line, causing a reflection voltage to be created. This voltage will add to the NEXT voltage calculated earlier, with the sum appearing as the total NEXT voltage, V_{fe}.

From elementary transmission line theory (see, for example, [4, 21–24]), the NEXT voltage V_{fe} will be equal to (8.31):

$$V_{fe} = NEXT(1 + \rho_L) \tag{8.31}$$

Applying (8.19), $\rho_L = +0.0604$.

Using (8.31), V_{fe} is then (178 mV × 1.0604) = 189mV. This matches the SPICE results of 190 mV in Figure 8.20.

Because $K_f = 0$, from (8.29), FEXT is zero. This follows directly from the discussion of the crosstalk model in stripline, which showed the inductively and capacitively coupled currents canceling at the far end. However, the simulation results displayed in Figure 8.20 show FEXT to be about 48 mV. This must mean that i_L and i_C did not cancel as expected, and the fact that the FEXT pulse is positive means $i_C > i_L$.

Three factors are responsible for the discrepancy between the simple hand calculation and the lossy line computer simulation:

- This line is lossy, and the simple equations presented here assumed a lossless line. The series resistance caused by the skin effect creates a larger $\dfrac{dv}{dt}$ to appear across each capacitor in Figure 8.21, increasing i_c without a corresponding increase in i_L.

- Just as in the NEXT case, the line's far end is mismatched by the 63-Ω load. This gives rise to a reflection to the voltage induced by the residual component of i_c across the 63-Ω resistor present at the far end, increasing the value of FEXT.

- Numerical inaccuracies are in the lossy SPICE transmission line model. The problem in this example is especially taxing because the simulator is required to manage very small numbers over many calculation iterations. In the simulator used to produce the results for this example, about 2 mV of the error can be attributed to this cause. Other simulators produce different error voltages.

8.5.7　Crosstalk Summary

As we've seen, coupling between conductors gives rise to crosstalk, so modifying the PWB design to reduce coupling will help improve crosstalk. The obvious first factor is the spacing s between traces, but the interplay between the trace width, thickness and ε_r all factor into determining the degree of coupling. The way these things interact is summarized below, in Table 8.4 for s, t, w, and ε_r as defined in Figure 8.25.

For example, for a given impedance, trace thickness and width, and dielectric constant, the first table entry shows (not surprisingly) that increasing s causes crosstalk to fall.

The next entry shows that when holding impedance, trace thickness, and ε_r constant, increasing the trace width (say from 4 mils wide to 8 mils wide) also decreases crosstalk for a constant $\frac{s}{w}$. This effect is more pronounced for thicker traces and is less evident in the thin traces used in micropackages. This means a 60-Ω 4-mil-wide half-ounce stripline having 4-mils edge-to-edge spacing would have higher coupling than a 60-Ω 8-mil half-ounce stripline on an 8-mil spacing. The severity of the change depends on the value of ε_r and Z_o (and so, h), but to cite one specific example K_b falls by over 10% in going from 4-mil-wide stripline with $s = 4$ to 8-mil-wide stripline having $s = 8$ for a 60-Ω half-ounce stripline on FR4. The discussion in Section 8.2.2 describes why increasing h causes k in increase.

8.6　Summary

Electric and magnetic coupling between conductors causes the impedance and time of flight to vary as neighboring signals switch, leading to data-dependent jitter. This

Table 8.4　Relationship Between PWB Physical Characteristics and Crosstalk

Holding Constant	Increasing	Causes Crosstalk to	Example
Z_o, t, w, ε_r	$\dfrac{s}{w}$	↓	Increasing s from 5 to 10 mils
Z_o, t, ε_r, $\dfrac{s}{w}$	ω	↓	4-mil-wide trace increased to 8 mil
t, w, ε_r, $\dfrac{s}{w}$	Z_o	↑	From 50-Ω to 65-Ω trace
Z_o, w, ε_r, $\dfrac{s}{w}$	t	↑	Half-ounce increased to one-ounce trace
Z_o, t, w, $\dfrac{s}{w}$	ε_r	↑	Going from a low ε_r laminate to FR4

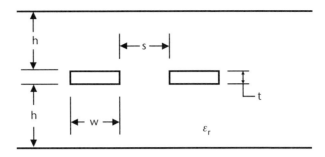

Figure 8.25 Stripline dimensions used in Table 8.3.

effect is readily modeled in SPICE-type simulators, and it's not necessary to include all of the conductors of a wide bus to obtain accurate results. Coupling falls off with distance, especially for low-impedance lines, so that only a few signals on either side of a victim need be included.

The switching activity can be categorized as either odd- or even-mode behavior. The way mutual capacitance and inductance is summed depends if the neighboring traces are switching in phase (even mode) or out of phase (odd mode).

Coupling also causes voltages to be induced from one or more culprit lines to one or more unswitching (passive) victim lines. The crosstalk voltages appearing at the line's near end are called NEXT; the far end voltages are called FEXT.

Simple equations for predicting NEXT and FEXT are useful for lossless lines but are somewhat more prone to error as losses increase, especially for FEXT. Multiconductor transmission line modeling using SPICE-type simulators that includes skin effect resistance provide accurate results, especially if the loads present at the far and near ends are well modeled.

In stripline $k_C = k_L$, which in lossless lines means that FEXT will be zero. It's nonzero in lossy lines, and the polarity depends on the relative magnitudes of the induced capacitive and inductive currents.

References

[1]   Cohn, S. B., "Shielded Coupled-Strip Transmission Line," *IRE Transactions on Microwave Theory and Techniques*, October 1955, pp. 29–38.

[2]   Matick, R., *Transmission Line for Digital and Communication Networks,* McGraw-Hill: New York, 1969.

[3]   Montrose, Mark I., *Printed Circuit Board Design Techniques for EMC Compliance*, New York: IEEE Press, 1996.

[4]   Bakoglu, H. B., *Circuits, Interconnections, and Packaging for VLSI*, Reading, MA: Addison-Wesley, 1990.

[5]   Paul, C. R., *Analysis of Multiconductor Transmission Lines*, New York: John Wiley & Sons, 1994.

[6]   Marx, K. D., "Propagation Modes, Equivalent Circuits, and Characteristic Terminations for Multiconductor Transmission Lines with Inhomogeneous Dielectrics," *IEEE Trans Microwave Theory and Techniques*, Vol. MTT-21, No. 7, July 1973, pp. 450–457.

[7]   "Interface Standard for Nominal 3V/3.3V Digital Integrated Circuits," JESD8-A, Electronics Industries Association, June 1994.

[8]   Skilling, H., *Electric Networks*, New York: John Wiley & Sons.

[9]   Electronic Industries Alliance, "EIA/JEDEC Standard for Operating Voltages and Interface Levels for Low Voltage Emitter-Coupled Logic (ECL) Integrated Circuits," EIA/JESD8-2, Arlington, VA, March 1993.

[10]   Federal Telephone and Radio Corp, *Reference Data for Radio Engineers*, 3rd Ed., New York: Knickerbocker Printing Corp., 1949.

[11]   Feller, A., et al, "Crosstalk and Reflections in High-Speed Digital Systems," *Proceedings of the Fall Joint Computer Conference*, Washington, D.C., December 1965, pp. 511–515.

[12]   Hart, B. L, *Digital Signal Transmission Line Circuit Technology*, Oxford: Van Nostrand Reinhold (UK), 1988.

[13]   Connolly, J. B., "Cross Coupling in High Speed Digital Systems," *IEEE Trans on Electronic Computers*, EC-15, No. 3, June 1966, pp. 323–327.

[14]   Catt, I., "Crosstalk (Noise) in Digital Systems," *IEEE Trans on Electronic Computers*, EC-16, No. 6, December 1967, pp. 743–763.

[15]   Rainal, A. J., "Transmission Properties of Various Styles of Printed Wiring Boards," *Bell System Technical Journal*, Vol. 58, No. 5, May–June 1979, pp. 995–1025.

[16]   DeFalco, J. A., "Reflections and Crosstalk in Logic Interconnections," *IEEE Spectrum*, July 1970, pp. 44–50.

[17]   Voranantakul, S., et al., "Crosstalk Analysis for High-Speed Pulse Propagation in Lossy Electrical Interconnections," *IEEE Trans. Components, Hybrids and Manufacturing Technology*, Vol. 16, No. 1, February 1993, pp. 127–136.

[18]   Kim, J., and J. F. McDonald, "Transient and Crosstalk Analysis of Slightly Lossy Interconnection Lines for Wafer Scale Integration and Wafer Scale Hybrid Packaging—Weak Coupling Case," *IEEE Trans. Circuits and Systems*, Vol. 35, No. 11, November 1988, pp. 1369–1382.

[19]   Gopinath, G., "Losses in Coplanar Waveguides," *IEEE Trans. Microwave Theory and Techniques*, Vol. MTT-30, No. 7, July 1982, pp. 1101–1104.

[20]   Knorr, J. B., et al., "Analysis of Coupled Slots and Coplanar Strips on Dielectric Substrate," *IEEE Trans. Microwave Theory and Techniques*, Vol. MTT-23, No. 7, July 1975, p. 541.

[21]   Johnson, W. C., *Transmission Lines and Networks*, New York: McGraw Hill, 1950.

[22]   Skilling, H. H., *Transient Electric Currents*, New York: McGraw Hill, 1952.

[23]   Sinnema, W., *Electronic Transmission Technology*, 2nd Ed., Englewood Cliffs, NJ: Prentice Hall, 1988.

[24]   Rosenstark, S., *Transmission Lines in Computer Engineering*, New York: McGraw Hill, 1994.

CHAPTER 9

Characteristics of Printed Wiring Stripline and Microstrips

9.1 Introduction

Previous chapters have discussed transmission lines in a general way, with little regard to their physical construction on a PWB.

This chapter discusses the specific electrical characteristics of stripline and microstrips and presents some simple formulas to calculate their loss, impedance, and time of flight. The literature abounds in formulas of varying complexity and accuracy to compute these parameters. Most were created before computer programs (*field-solving software*) capable of calculating the RLGC parameters from the transmission line's physical construction were widely available. The more complex equations generally have better accuracy, but the simpler ones presented here are accurate enough for estimation purposes. They are included in this chapter as a way to illustrate the underlying principles.

The chapter concludes with a discussion of forming differential pairs using various types of microstrip and stripline.

9.2 Stripline

Stripline trace is formed when a trace is immersed in a sea of dielectric and is sandwiched between two return planes. The construction is called *symmetrical stripline* (usually simply *stripline*) when the trace is centered in the dielectric so that its top and bottom are the same distance away from their respective return planes. This is shown in Figure 9.1.

Placing the trace closer to one plane than the other forms *offset stripline* (also usually called *stripline*).

Although the physical construction differs, in both forms of stripline the magnetic and electric field lines are confined to the dielectric in the space between the return planes. Thus, true TEM propagation occurs if the losses are small. This is in contrast with microstrip (discussed in Section 9.3), where some of the field lines propagate in air, complicating the creation of simple impedance and time-of-flight formulas. Stripline's homogeneous dielectric makes these calculations straightforward, with the only complicating factor being the proper accounting of fringing at the trace's ends during calculation impedance.

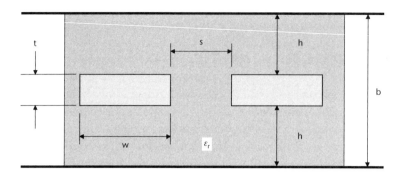

Figure 9.1   Stripline edge view.

9.2.1   Time of Flight

From Maxwell's equations (for example, see [1]), TEM waves traveling in a homogeneous dielectric (such as stripline) propagate at a velocity equal to (9.1):

$$V_p = \frac{1}{\sqrt{\mu_o \mu_r \varepsilon_o \varepsilon_r}} \tag{9.1}$$

As defined in Chapters 3 and 4, ε_o is the permittivity of free space ($8.854 10^{-12}$ F/m $\cong 224.9$ fF/in), ε_r is the relative permittivity (see Section 3.2.1), μ_o is the permeability of free space ($4\pi \times 10^{-7}$ H/m, or about 31.9 nH/in), and μ_r is the relative permeability (equal to one for nonferris materials such as copper—see Section 4.2.1).

In a vacuum, μ_r and ε_r are both equal to one, and V_p becomes the constant c, the speed of light in a vacuum:

$$V_p = \frac{1}{\sqrt{\mu_o \varepsilon_o}} = \frac{1}{\sqrt{4\pi \times 10^{-7} \times 8.85 \times 10^{-12}}} \equiv c = 3 \times 10^8 \, \text{m/s} \tag{9.2}$$

Using the results from (9.2) as a constant rather than solving (9.1) each time is convenient when $\varepsilon_r \neq 1$. Assuming $\mu_r = 1$, (9.1) simplifies to (9.3):

$$V_p = \frac{c}{\sqrt{\varepsilon_r}} \tag{9.3}$$

Putting it into words, (9.3) shows that in nonferric materials (where $\mu_r = 1$), the dielectric will slow the propagation of TEM waves (such as those traveling along stripline traces) by $\sqrt{\varepsilon_r}$ relative to the speed of light in a vacuum. This is independent of the structure's geometry. At the circuit-board level this means that, provided ε_r is truly the same, on a given board all striplines, regardless of their impedance, width, or thickness will propagate signals at the same velocity.

Delay per unit length is the inverse of velocity, which leads to the following for the time of flight along striplines (this was offered without proof in Section 5.4.2) (9.4):

$$t_d = \frac{\sqrt{\varepsilon_r}}{c} \tag{9.4}$$

Using engineering units, (9.4) may be rewritten as (9.5):

$$t_d = 84.72\sqrt{\varepsilon_r}\,\text{ps/in} \qquad (9.5)$$

Which is equivalent to 1.017 ns/ft when $\varepsilon_r = 1$.

Because in TEM propagation, μ is analogous to L and ε is analogous to C, (9.1) may be written as (9.6) (for example, see [1, 2]):

$$V_p = \frac{1}{\sqrt{LC}} \qquad (9.6)$$

Which leads to (9.7), to the delay per unit length for a TEM transmission line:

$$t_d = \sqrt{LC} \qquad (9.7)$$

This is the same result as (5.17), appearing in Section 5.4.3.

Example 9.1

Two striplines are fabricated on the same piece of FR4 ($\varepsilon_r = 4.5$). The first is made with $h = 5.4$ mils and has an inductance $L = 8.08$ nH/in and a capacitance $C = 4$ pF/in. The second stripline is spaced twice as far from the reference planes: $h = 10.8$ mils, yielding $L = 11.3$ nH/in and $C = 2.84$ pF/in.

Use (9.4) and (9.7) to compute the delay per inch for both cases.

Solution

(a) From (9.7), $t_d = \sqrt{LC} = \sqrt{8.08\ \text{nH} \times 4\ \text{pF}} = 179.8$ ps/in for the $h = 5.4$ mils case, and $t_d = \sqrt{LC} = \sqrt{11.3\ \text{nH} \times 2.84\ \text{pF}} = 179.8$ ps/in for the $h = 10.8$ mils case.

As this line is a simple stripline, the trace is immersed in a single dielectric and the delay per inch is unrelated to the spacing h. This comes about because for a given width doubling the spacing causes the capacitance to be reduced, while the inductance is simultaneously increased by a compensating amount. This is due to the reciprocity principal described in Section 4.2.5. The net result is that t_d is identical for both striplines. As an aside, since microstrip doesn't truly propagate TEM those types traces won't behave in this way.

(b) From (9.4), $t_d = \dfrac{\sqrt{\varepsilon_r}}{c} = \dfrac{\sqrt{4.5}}{11.8 \times 10^9\ \text{in/s}} = 179.8$ ps/in for both cases.

Notice that t_d changes as $\sqrt{\varepsilon_r}$ rather than linearly, so t_d moves somewhat slowly with changes in ε_r. For example, changing ε_r by 11% from 4.5 to 4.0 results in t_d being reduced from 179.8 pS to 165.9 pS (a 5.7% reduction).

9.2.2 Impedance Relationship Between Trace Width, Thickness, and Plate Spacing

It was seen in Chapter 5 that a trace's impedance is inversely proportional to the square root of its capacitance (9.8):

$$Z_o = \sqrt{\frac{L}{C}} \qquad (9.8)$$

In Chapter 3, capacitance was shown to increase directly with the relative dielectric constant ε_r. Therefore, from (9.8) for a given w and b, impedance will fall as the square root of the increase in ε_r.

For example, if the impedance for a particular stripline fabricated on a laminate having $\varepsilon_r = 3.5$ is 65Ω, it will fall to $65\sqrt{\dfrac{3.5}{4.5}} = 57.3\Omega$ when fabricated on a laminate having $\varepsilon_r = 4.5$.

Chapter 3 also showed that capacitance increases as the trace width w increases, but decreases as the separation between plates grows larger. Neglecting fringing, this suggests that a given stripline capacitance (and by implication, Z_o) can be obtained by holding the ratio of $\dfrac{w}{b}$ constant, and this is indeed the case. In fact, accounting for trace thickness t, the relationship between $\dfrac{w}{b-t}$ and impedance is nearly independent of trace width and thickness, especially for a given thickness and within a small range of changes in width. This is evident in Figure 9.2, which shows impedance plotted against $\dfrac{w}{b-t}$ for three values of ε_r. The graph is for illustration purposes only and is not intended as a general-purpose design tool. It was created from field-solving software for $t = 0.65$ mils and $w = 5$ mils and so is most accurate under these and similar conditions.

From Figure 9.2, a $\dfrac{w}{b-t}$ is seen to result in a 53-Ω trace if $\varepsilon_r = 4.5$. This means for a 5-mil-wide, 0.65-mil-thick trace, the required plate spacing b is about 15 mils.

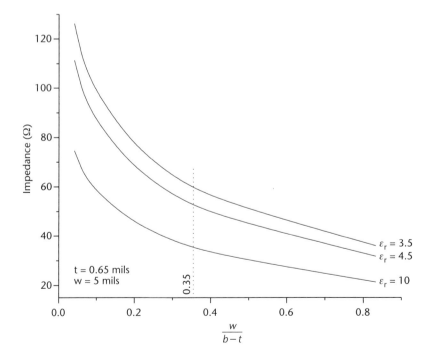

Figure 9.2 Z_o versus $\dfrac{w}{b-t}$ for stripline.

Increasing the trace width to 10 mils requires b to be increased to 29.2 mils in order to hold $\dfrac{w}{b-t}$ constant and thus to obtain the same 53-Ω impedance. Holding the dimensions constant but reducing ε_r to 3.5 raises the impedance to about 60Ω.

9.2.3 Mask Biasing to Obtain a Specific Impedance

As a practical matter, manufacturers must choose between specifically available laminate thicknesses when they fabricate a PWB. Many different thicknesses (i.e., $\dfrac{b-t}{2}$) are available from stock for common laminates such as FR4, but often the selection is limited for some of the more exotic laminates. In those situations where the specific calculated value is not available from stock, manufacturers can take advantage of the relationship described in Figure 9.2 to photo lithographically adjust w to obtain the desired impedance. In this case, the trace width printed on the PWB is different (usually smaller) than that appearing in the design database artwork. This is sometimes called *biasing the artwork* or *biasing the trace* and is more likely to occur if the trace has been specified as *controlled impedance*. Narrowing the trace increases conductor loss (especially at high frequency), so in situations where loss is critical it's important for the designer to determine if the printed trace widths match those appearing in the artwork.

9.2.4 Hand Calculation of Z_o

Calculating the impedance of a single stripline trace is fairly straightforward if the trace has zero thickness, but gets more complicated as its thickness increases and fringing becomes significant.

Numerous models have appeared in the literature for hand-calculating stripline trace impedance, with Cohn's [3, 4] among the first and probably the most widely known. All of these models assume a rectangular-shaped trace and stripline centered between the reference planes. As was seen in Chapter 1, etching makes most traces trapezoidal, with the most severe undercutting likely to appear on thick traces. Undercutting effects tend to reduce a trace's capacitance (and therefore increase its impedance).

The model presented by Cohn requires the trace to be no thicker than 25% of the plate spacing. For a half-ounce trace this means $b \geq 2.6$ mils (or 5.6 mils for 1-oz trace), which is generally met in practical PWB designs, especially if w and ε_r are not too small. The trace is also required to be wide with respect to b and t such that $\dfrac{w}{b-t} \geq 0.35$.

To calculate the impedance under these conditions, Cohn developed the following, here recast in a simpler form similar to that presented in [5]:

$$Z_o = \frac{94.15}{\sqrt{\varepsilon_r}\left(\dfrac{w}{b}k + \dfrac{C_f}{8.854\varepsilon_r}\right)} \tag{9.9}$$

where

$$k \equiv \frac{1}{1 - \dfrac{t}{b}}$$

(9.10)

The fringing capacitance C_f is given as:

$$C_f = \frac{8.854\varepsilon_r}{\pi} \left[2k\ln(k+1) - (k-1)\ln(k^2-1) \right] \text{ in pF/m}$$

(9.11)

Although C_f is in pF/m, (9.9) has been scaled by using $\varepsilon_o = 8.854$ pF/m in both (9.9) and (9.11). This permits (9.9) to accept the results from (9.11) directly, so that either metric or common engineering (inch or mil based) units may be used when specifying w, b, and t in (9.9) and (9.10).

Figure 9.2 shows that (9.9) is most accurate for a trace impedance under about 65Ω for most practical PWB laminates, as that's where $\dfrac{w}{b-t} \geq 0.35$. Under these conditions, (9.9) slightly (to within ~ 2%) underestimates the impedance when compared to results generated by field-solving software (e.g., [6]) at a specific frequency where ε_r is known. The accuracy improves to better than 1% for larger values of $\dfrac{w}{b-t}$.

A different model more suitable for higher impedance trace where $\dfrac{w}{b-t} \leq 0.35$ is presented in [7] and appears in (9.12):

$$Z_0 = \frac{60}{\sqrt{\varepsilon_r}} \ln\left[\frac{4b}{\pi d} \right]$$

(9.12)

where d is defined as (9.13):

$$d = \frac{w}{2} \left[1 + \frac{t\left[1 + \ln\left(\dfrac{4\pi w}{t} \right) + 0.51\pi\left(\dfrac{t}{w} \right)^2 \right]}{\pi w} \right]$$

(9.13)

Combined, these equations give results matching to better than 2% of those obtained by field-solving software across a wide range of impedances and trace widths. They are easily programmed in numerical software such as Mathcad [8] or MATLAB [9] incorporating a test on $\dfrac{w}{b-t}$ to determine the equation set to use.

Example 9.2

Compute Z_o for a half-ounce (t = 0.65 mils), 5-mil-wide stripline on FR4 ($\varepsilon_r = 4.5$) when (a) $b = 6$ mils and (b) $b = 24$ mils. Assume the traces are perfectly rectangular.

Solution

(a) With $b = 6$ mils, $\dfrac{w}{b-t} = 0.417$, so (9.9) applies.

First solving (9.11) yields $C_f = 20.757$ pF/m. Even though w and b are in mils, C_f is used as is in (9.9). Solving (9.9) yields $Zo = 47.34\Omega$, versus 48Ω from field-solving software (1.4% low).

(b) With $b = 24$ mils, $\dfrac{w}{b-t} = 0.104$, making (9.12) applicable. Solving (9.13) yields $d = 3.079$. Applying that to (9.12) produces $Z_o = 84.9\Omega$ versus 85.2Ω from field-solving software (0.4% low).

9.2.5   Stripline Fabrication

Implied in (9.2) and Figure 9.1 is that the stripline trace is embedded in a sea of dielectric that has the same value everywhere. In practice, microstrips are not formed in quite this way. Instead, two very similar but distinct dielectrics are involved: the copper-cladded laminate sheet and the prepreg used to bond together adjacent sheets. This is shown in Figure 9.3.

To form stripline, the PCB fabricator receives the laminate sheets from the manufacturer coated with cooper foil on both sides and etches the signal traces on one of the sides. Multiple sheets are bonded together with prepreg to form the final

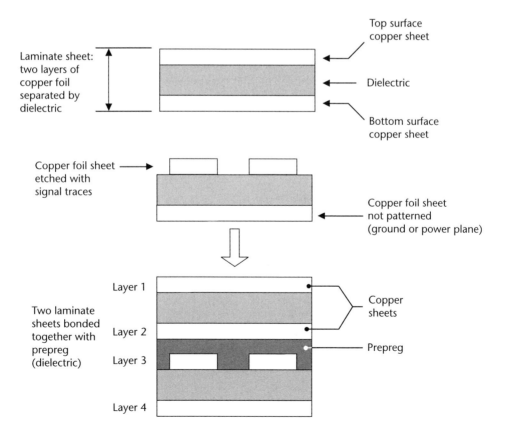

Figure 9.3   Stripline construction.

stackup. Four layers are shown in Figure 9.3, with one sheet comprising layers 1 and 2 and a second sheet forming layers 3 and 4. Layer 3 is a signal layer, creating a stripline with layers 2 and 4. The stripline is thus formed from two separate laminate sheets. Although in Figure 9.3 L1 is shown as a plane, it could be patterned with signal traces.

As discussed in Chapter 1, the laminate and prepreg ε_r values are affected by their thickness and by the glass-to-resin ratio (see Tables 1.2 and 3.2 and Figure 1.2). A weighted average between the two determines the final value of ε_r for that layer. This means that in general ε_r will not be the same for the prepreg and laminate, especially in FR4 boards. This is illustrated in Table 9.1, which shows the thickness in mils and the corresponding ε_r value for three separate circuit board designs using a high-TG FR4-type laminate system.

Cases 1 and 2 are the more typical situations, where the difference in ε_r between the laminate and the prepreg is well under 10%, but case 3 illustrates a situation where the manufacturer opted to use a thick application of resin with little reinforcing glass to bond to a core that also had a relatively low glass content. The intent was to drive the aggregate ε_r lower than typically found on FR4 boards, but doing so resulted in a difference of greater than 14% in the dielectric constants of the two materials. Proper accounting of these construction details is necessary when modeling the impedance, time of flight, and loss (the latter especially so at very high signaling rates).

Usually fabricators select the laminate and prepreg thickness on a layer-by-layer basis to help obtain a desired impedance and overall board thickness. From (9.4), it follows that doing so will affect the time of flight, as ε_r will change with thickness. Layer-by-layer thickness adjustment is more prevalent on thick, multilayer boards and is one reason why various routing layers intended to be identical can have different times of flight on the same PWB. It's also a factor in why the same PWB artwork can perform differently when it's made by multiple vendors using identical materials. Along with biasing the trace to obtain a specified impedance, vendors will use unique laminate and prepreg combinations that historically have yielded well for them. Unless the identical stackup is used, this results in ostensibly identical boards obtained from different vendors having different times of flight.

These construction details can have consequence when forming broadside coupled differential pairs and delay lines. It's also necessary to understand these construction details to develop accurate two-dimensional or three-dimensional field-solver models.

Table 9.1 Differences in Hi-TG FR4 Laminate and Prepreg ε_r			
Case	Prepreg Thickness (mils)/ε_r	Core Thickness (mils)/ε_r	Change ε_r
1	4.8/4.36	5.2/4.60	5.5%
2	4.8/4.36	22/4.65	6.7%
3	4.7/3.65	3.8/4.19	14.7%

9.3 Microstrip

In contrast with stripline, in microstrip the trace is referenced to a single return plane and is separate from it by a slab of dielectric, as shown in Figure 9.4. Because the trace is also exposed to the air, fields propagate along microstrip in two very different dielectrics.

Surface finishes are put on the microstrips to prevent corrosion. As described in Chapter 1, most of these finishes are conducting (such as HASL, which is solder, or immersion silver, which is an organic material impregnated with silver particles). Often fabricators will *plate up* the surface copper from half- or one-ounce copper thickness to something higher in order to achieve an overall board thickness or as a way to obtain the desired impedance. It's not uncommon for what was specified as half-ounce microstrip to actually appear on the board's surface two to four times thicker than that due to plating. The plating and etching processes ordinarily make the trace shape nonrectangular.

To aid in the soldering and assembly process, fabricators typically put a thin (usually under 1 mil thick) coating of epoxy everywhere on the board except where connections are to be made (such as component solder pads and gold-plated fingers for edge connectors). This *solder mask* prevents solder from flowing to where it's not wanted, and effectively embeds the microstrip in a thin second dielectric (see Figure 9.5). In this configuration, fields propagate in three dielectrics (the laminate, solder mask, and air).

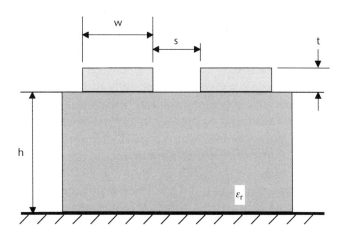

Figure 9.4 Microstrip edge view.

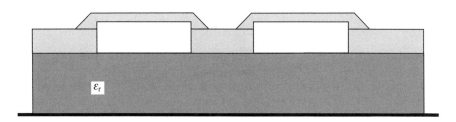

Figure 9.5 Microstrip coated with solder mask.

The use of surface finishes and solder mask on microstrip greatly complicates their modeling, especially at high frequencies where the skin effect causes the signal currents to migrate to the conductor's surface and where the lossy nature of the solder mask contributes to the overall dielectric loss.

If the board is designed appropriately, solder mask can be specifically excluded on specific regions of a board (or the entire board). This eliminates the solder mask dielectric loss on those traces where total loss is critical and can be especially attractive when a low ε_r, low loss laminate is used. This is addressed in Section 9.3.2.

9.3.1 Exposed Microstrip

In microstrip not covered by solder mask, some of the field lines travel in the dielectric and some in the air, with the concentration depending on the precise values of ε_r and the $\dfrac{w}{h}$ ratio. Incorporating these effects makes hand calculation of time of flight and impedance not as simple as for stripline. Many researchers (for example, see [10–14]) have developed closed-form microstrip impedance equations, with the usual approach being to treat the trace as if it were in a single, uniform dielectric having a dielectric constant somewhere between air ($\varepsilon_r = 1$) and the value of the laminate. The *effective dielectric constant*, ε_{r_eff}, is calculated first and then used in a related companion formula to calculate Z_o or directly used in (9.4) to determine the time of flight. Because the dielectric is not homogeneous, microstrips do not truly propagate TEM. However, by using the effective dielectric constant approach, they can be modeled as if they do, especially if the losses are small. Microstrips are thus said to propagate *quasi-TEM*.

Some of the microstrip impedance equations are quite involved, but even still their results only come to within about 10% of field-solving software for narrow trace on low dielectric constant laminates. Fortunately, accuracy improves with the wider type trace fabricated on higher ε_r laminates, such as those in use on most PWB designs. In this situation, the error is typically within a few percent of that obtained by field-solving software. None of the simple equations address the case when solder mask is present.

The following two equations give good results and are not too complex. With the availability of field-solving software, equations of greater complexity are not worth the required extra effort. As was the case in stripline, accuracy is improved by using a test based on the ratio of trace width and height to select between the regimes of operation modeled by the two equations.

From [10, 11], use (9.14) when $\dfrac{w}{h} < 2$ (on FR4 this represents $Z_o \sim 60\Omega$ and higher without solder mask on FR4 for typical trace widths):

$$Z_o = \frac{60}{\sqrt{\varepsilon_{r\_eff}}} \ln\left(\frac{5.98h}{0.8w + t}\right) \qquad (9.14)$$

where ε_{r_eff} is defined in (9.15):

$$\varepsilon_{r\_eff} = 0.475\varepsilon_r + 0.67 \qquad (9.15)$$

From [13], use (9.16) when $\dfrac{w}{h} \geq 2$ (corresponding to an impedance under about 60Ω on FR4):

$$Z_o = \frac{120\pi}{\sqrt{\varepsilon_{r\_eff}}\left[\dfrac{w}{h} + 1.393 + 0.667\ln\left(\dfrac{w}{h} + 1.444\right)\right]} \qquad (9.16)$$

Equation (9.16) works best when ε_{r_eff}, as defined in [14] is used (9.17):

$$\varepsilon_{r\_eff} = \frac{\varepsilon_r + 1}{2} + \left(\frac{\varepsilon_r - 1}{2}\frac{1}{\sqrt{1 + \dfrac{10h}{w}}}\right) \qquad (9.17)$$

Example 9.3

Compute Z_o and t_d for a half-ounce microstrip with $h = 3.5$ mils on a laminate having $\varepsilon_r = 4.0$ when (a) $w = 4$ mils and (b) when w is increased to 8 mils. Assume the trace to be rectangular.

Solution

(a) With $h = 3.5$ mils and $w = 4$ mils, $\dfrac{w}{h} = 1.1$, making (9.14) applicable. Solving (9.15) yields $\varepsilon_{r\_eff} = 2.57$. Using that in (9.14) yields $Z_o = 63.4\Omega$, versus 65.9Ω from field-solving software (3.8% low).

    With $\varepsilon_{r\_eff} = 2.57$, (9.4) yields $t_d = 135.7$ pS/in.

(b) With $w = 8$ mils, $\dfrac{w}{h} = 1.1$, making (9.16) applicable. Solving (9.17) yields $\varepsilon_{r\_eff} = 3.147$. Applying that to (9.16) produces $Z_o = 46.6\Omega$ versus 45.6Ω from field-solving software (about 2.3% high).

    With $\varepsilon_{r\_eff} = 3.147$, (9.4) yields $t_d = 150.2$ pS/in.

The time of flight results are of particular interest. Because the fields are propagating in both air and laminate, the signals should propagate faster than if just in the laminate but slower than if just in air. It follows from (9.5) that in air (where $\varepsilon_r = 1$) $t_d \sim 85$ pS/in, and from the time of flight calculations, it's apparent that in the $h = 6$ mils case, more of the fields propagate in the air than in the $h = 1.5$ mils case. In fact, in this particular example, increasing the height from 1.5 mils to 6 mils results in a higher impedance trace having a nearly 4.5% smaller delay. This makes intuitive sense: being further away from the return reduces the trace capacitance (and thus raises its impedance), and lowers the concentration of field lines in the dielectric verses those in the air. As more of the fields are propagating in the air, the trace should be "faster" than when the trace is closer to the return, as Example 9.3 shows.

9.3.2 Solder Mask and Embedded Microstrip

Usually high-speed logic circuit boards will be covered in solder mask, creating an embedded microstrip, as was shown in Figure 9.5. The solder mask chemistry and final thickness when dry varies between fabricators and PWB topology, but from experience a typical application is a 0.6–0.8-mil-thick coating over the copper with $\varepsilon_r \sim$ 3.1–3.3. A nominal application is 0.7 mils thick with ε_r = 3.2. Loss tangent data is not readily available, but a value close to that of the epoxy resin used in FR4 (i.e., $\tan\delta$ = 0.020) is a good approximation.

Embedding the trace in solder mask makes the impedance partially a function of its thickness. As three dielectrics are involved, the prediction of ε_{r_eff} (and thus of Z_o and t_d) becomes more involved. This difficult problem is best analyzed with field-solving software, but some researchers have produced analytical results [15–17]. The equations are complex and so tend to inhibit an intuitive insight as to the effects solder mask will have on signal propagation. A qualitative argument and results from field-solving software will be used here instead.

Example 9.3 showed that the proportion of field lines propagating in the air versus those in the laminate determined t_d and Z_o. The trace was faster if more lines were propagating in the air (ε_r = 1) than in the laminate (ε_r > 1). As solder mask covers the trace with a dielectric having ε_r > air, microstrips covered with solder mask will be somewhat slower than the exposed microstrip in Example 9.3. The actual propagation speed will depend on the relationship between the solder mask and laminates ε_r and the values for h, w, and t. This is because they all interplay to determine the field concentration in each of the three dielectrics, as is shown in Table 9.2. The results are from field-solving software for half-ounce copper of two widths (w = 3 mils and w = 8 mils) at a frequency of 1.25 GHz.

The b subscript in the table denotes bare board (void of solder mask); the s subscript shows results with an application of 0.7-mil-thick solder mask on top of the copper, with ε_r = 3.2 and tan = 0.02. In both cases, the board is FR4 (ε_r = 4.5, $\tan\delta$ = 0.02). The values for α_{t_b} and α_{t_s} are for the combined dielectric and conductor losses at 1.25 GHz.

In the narrow trace, lower impedance case, the impedance is seen to change from 49Ω to 44Ω (~10%) with the application of solder mask, while the wide trace is less affected (~ 4% in going from 47Ω to 45Ω). As expected, with less of the field lines in the dielectric, increasing h makes the trace faster, even with the application of solder mask.

If the microstrip is specified as *controlled impedance*, the vendor will adjust the physical layup of the PWB to obtain the required impedance in the presence of the

Table 9.2 Effects of Solder Mask on Microstrip Z_o, t_d, α_t

w (mils)	w/h	Z_{-b} (Ω)	Z_{-s} (Ω)	t_{d_b} (pS/in)	t_{d_s} (pS/in)	αt_{-b} (dB/in)	at_{-s} (dB/in)
3	2	49	44	151	166	0.30	0.34
3	0.5	89	83	147	158	0.18	0.20
8	2	47	45	155	163	0.19	0.21
8	0.5	92	89	148	154	0.13	0.14

solder mask. Adjusting w or h (either alone or in combination) will yield the proper impedance, but these parameters have separate effects on t_d and α_r. Reducing w (*biasing the trace*, as described in Section 9.2.3) in the presence of solder mask will increase Z_o and increase conductor loss but will reduce t_d. Increasing h increases Z_o and reduces the dielectric losses and the time of flight, even in the presence of solder mask.

Application of solder mask is seen to have the largest effect on α_t for the narrow trace close to the surface (increasing the loss by 6.25%). The impact diminishes as $\frac{w}{h}$ falls.

9.4 Losses in Stripline and Microstrip

Previous chapters discussed the general nature of transmission line losses, and showed how to compute them if the RLCG values are known. Usually for PWB transmission lines, the designer knows the line's impedance, time of flight, and physical dimensions instead of these elementary electrical parameters. Those values can be hand calculated or determined by field-solving software, but suitable closed-form formulas that explicitly calculate conductor and dielectric losses for stripline and microstrip trace are more convenient. Such formulas are presented in this section.

From the discussion of general lossy transmission lines given in Chapter 5, the total loss is the sum of conductor and dielectric losses (9.18):

$$\alpha_t = \alpha_c + \alpha_d \tag{9.18}$$

Conductor loss α_c is due to the resistive losses in the conductor and return path (see Chapter 2).

From Chapter 3, the dielectric loss α_d is determined by the value of the loss tangent and represents energy lost to the dielectric.

The construction of a 50-Ω stripline and solder-mask-covered microstrip appear in Figure 9.6.

The difference in construction suggests the relationship between dielectric and conductor losses would not be the same for stripline and microstrip. Intuitively, the two return planes in stripline should cause lower total loop conductor loss than would be present in microstrip, but dielectric losses should be higher because the fields are totally bound within a dielectric that is more lossy than air. This contrasts with microstrip, where only one plane is available to the return current (increasing the total loop resistance), and at least some of the fields propagate in air (which is lossless).

This intuitive analysis is confirmed in Figure 9.7. It shows losses for a 5-mil-wide half-ounce 50-Ω stripline and microstrip coated with solder mask on FR4 ($\varepsilon_r = 4.25$). The solder mask is 0.7 mils thick, with an ε_r of 3.2. The stripline is represented by the solid lines; microstrip is represented by the broken ones.

As expected, stripline's dielectric loss (α_d) is seen to be greater than that of microstrip, and microstrip's conductor losses (α_c) are higher. The net result is that

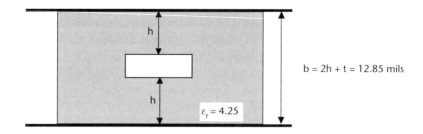

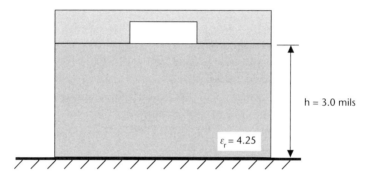

Figure 9.6   Dimensions of 50-Ω stripline and solder-mask-covered microstrip on FR4.

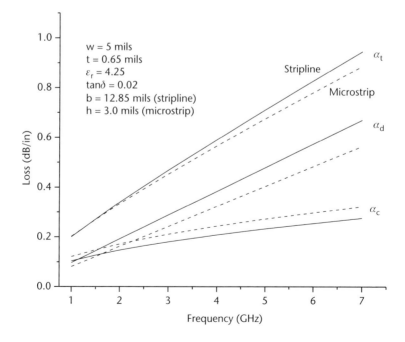

Figure 9.7   Losses in 50-Ω stripline (solid curve) and microstrip (broken curve). In this example, dielectric loss exceeds conductor loss for frequencies greater than 1GHz.

the total losses (α_t) are slightly higher for stripline than for microstrip over the frequencies appearing in the figure.

9.4.1   Dielectric Loss

Closed-form equations to estimate stripline conductor and dielectric losses are given in [4]. The dielectric loss formula is recast in (9.19) to present results in decibels per inch length:

$$\alpha_d = 2.318 f \sqrt{\varepsilon_r} \tan \delta \tag{9.19}$$

where f the frequency in gigahertz, ε_r the dielectric constant and $tan\delta$ the loss tangent. The accuracy of (9.19) is better than 1%.

Equation (9.19) may be used to approximate α_d in microstrip by replacing ε_r with ε_{r_eff}. Doing so for exposed microstrip gives results to better than 10% of those obtained by field solving software for small values of $\dfrac{w}{h}$ on FR4 ($\varepsilon r \sim 4.25$). The accuracy improves as $\dfrac{w}{h}$ increases.

9.4.2   Conductor Loss

In Chapter 5, α_t was shown to be approximately equal to (9.20):

$$\alpha_t \approx \frac{R}{2Z_0} + \frac{GZ_0}{2} \tag{9.20}$$

The first term is implied to be the conductor losses (α_c), and the second, the dielectric losses (α_d). Under the circumstances described in Chapter 5, the conductor losses portion is therefore (9.21):

$$\alpha_c = \frac{R}{2Z_0} \tag{9.21}$$

The equations presented in Chapter 2 to determine conductor and return path resistance at a specific frequency [(2.14) through (2.17)] may then be used to determine R. Although applicable to either microstrip or stripline, this process is tedious, and its accuracy depends on the ability to estimate the total loop resistance at frequency. Fortunately, simple formulas incorporating frequency-dependent losses suitable for hand calculation have been developed specifically for stripline and microstrip. The following formulas avoid the laborious process of first estimating the loop ac resistance as described in Chapter 2 and applying it to propagation constant in Chapter 5.

9.4.2.1   Stripline

The relationship developed in [4] for conductor losses is presented in (9.22) but recast for simplicity and formatted in engineering units, with results in decibels per inch length:

$$\alpha_c = \frac{2.02 \times 10^{-3} \varepsilon_r Z_o \sqrt{f}}{b} \left[k + \frac{\frac{2w}{b}}{\left(1 - \frac{t}{b}\right)^2} + \frac{1}{\pi} \frac{1 + \frac{t}{b}}{\left(1 - \frac{t}{b}\right)^2} \ln\left(\frac{k+1}{k-1}\right) \right] \qquad (9.22)$$

where b, w, and t are as shown in Figure 9.1 and have units of mils. The constant k is given in (9.10), and Z_o may be computed by (9.9) or could be the results obtained from field-solving software. The frequency f is in gigahertz.

Proper estimation of skin-effect resistance in the trace and return path is difficult to achieve in a simple equation, and (9.22) tends to underestimate the conductor loss as a result. Similar to the impedance formulas, the original work presented separate equations for *wide strip* and *narrow strip* traces. Equation (9.22) is a reformulation of the equation presented in [4] for wide strips and strictly speaking is only valid when $\frac{w}{b-t} \geq 0.35$ (i.e., wide trace having impedances below about 65Ω, according to Figure 9.2). However, analysis shows (9.22) has acceptable hand calculation accuracy for higher impedance trace. With the availability of field-solving software, the improvement obtained by the use of the more involved equation set is not worth the added complexity.

Example 9.4

What is α_t at 5 GHz for a 5-mil-wide, half-ounce ($t = 0.65$) stripline on FR4 ($\varepsilon_r = 4.25$) with $b = 36.65$ mils? From field-solving software, Z_o is known to be 80Ω.

Solution

The values for α_c and α_d must first be found to solve (9.18). Equations (9.22) and (9.10) are required to determine α_c; (9.19) is used to find α_d.

From (9.10), $k = 1.018$, and from (9.22) $\alpha_c = 0.12$ dB/in.

It's interesting to note that in this example $\frac{w}{t-b} = 0.139$, which is well outside the stated operating range for this equation. Nonetheless (9.22) has produced a result that's within 10% of that predicted by field-solving software.

From (9.19) $\alpha_d = 0.46$ dB/in, about 3% lower than expected from field-solving software.

Using (9.18) $\alpha_t = 0.12 + 0.46 = 0.58$ dB/in, which is about 5% lower than predicted by field solving software. This degree of accuracy in the overall loss comes about because in this example α_d is so much larger than α_c, and (9.19) predicts α_d with good accuracy. The results are not as good at lower frequencies, where α_c is close to or larger than α_d because in that situation the inaccuracies of (9.22) will dominate.

9.4.2.2 Microstrip

Many researchers have studied conductor loss in microstrip (for example, see [18–22]). The following equation (9.23) is recast from [19] and is suitable for $\frac{w}{b}$ ranging between 0.159 and 2 (i.e., exposed PWB microstrip on FR4 from roughly

50Ω to over 100Ω). Other equations are presented in [19] for wide, lower impedance traces.

Although the trace dimensions may be entered in mils, the units cancel and the results produced by (9.23) are in decibel per inch length. The frequency f is in gigahertz.

$$\alpha_c = \frac{11.411\sqrt{f}}{hZ_o}\left(\left(1-\left(\frac{w_p}{4h}\right)^2\right)\left(1+\frac{h}{w_p}+\frac{h}{\pi w_p}\left(\ln\left(\frac{2h}{t}\right)-\frac{t}{h}\right)\right)\right) \qquad (9.23)$$

where w_p is a width-correction factor based on trace thickness and height (9.24):

$$w_p = w + \left(\frac{t}{\pi}\left(\ln\left(\frac{2h}{t}\right)+1\right)\right) \qquad (9.24)$$

Example 9.5

Find the total loss at 5 GHz for the exposed microstrip described in part (a) of Example 9.3 ($h = 3.5$ mils, $w = 4$ mils, $t = 0.65$ mils, $\varepsilon_r = 4.0$) The results of that example showed the trace impedance to be 63.4Ω.

To find the total loss as represented in (9.18), the values for α_c and α_d must be found first. Equations (9.23) and (9.24) are required to determine α_c; (9.19) is used to find α_d.

From (9.15), ε_{r_eff}, and from (9.19) $\alpha_d = 0.372$ dB/in.

From (9.24), $w_p = 4.699$ mils. Using that in (9.23) yields $\alpha_c = 0.235$ dB/in.

Finally, using (9.18) the total loss is: $\alpha_t = \alpha_c + \alpha_d = 0.607$ dB/in (versus the field solver total of 0.62 dB/in).

9.5   Microstrip and Stripline Differential Pairs

The electrical aspects of traces forming differential pairs are discussed in Section 8.4, with no consideration for the way those pairs are actually created. Differential pairs can be formed either on the board's surface in a microstrip configuration, or deeper in the board, as stripline. Stripline differential traces may be formed as either edge-coupled or broadside pairs. The electrical characteristics of differential pairs fabricated on circuit boards are discussed next.

9.5.1   Broadside Coupled Stripline

Forming a broadside differential pair is illustrated in Figure 9.8. Four copper layers (L1 through L4) are used, with L1 and L4 being return planes, and layers L2 and L3 forming the differential pair signal carrying conductors.

A laminate sheet may be used to form layers (L1, L2) with a separate sheet forming layers (L3, L4). Prepreg would be used between layers L2 and L3. Alternatively, layers (L2, L3) may be formed on a sheet with layers L1 and L4 on two additional sheets. In this case prepreg fills the void between (L1, L2) and between (L2, L4). The fabricator usually determines these construction details. Generally L2 and L3 will

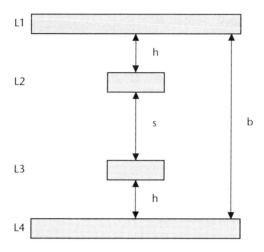

Figure 9.8 Broadside differential pair.

be formed on a single laminate sheet if the design requires tight control on the spacing between these layers, or if the alignment between the L2 trace and L3 trace is important (as is often the case). Alternatively, if *h* is more critical than *s*, or if the design can tolerate L2/L3 misalignment, the fabricator may opt to form (L1, L2) and (L3, L4) on laminate sheets and use prepreg to fill the void between them.

With the proper connector system, it's trivial to obtain precise trace length matching between the two signals forming the pair. In Figure 9.8, the routing on layer L2 is simply copied to layer L3 to form the diff pair. The only difference in the trace length is a small amount at the very ends of the traces where they connect to the connector pins.

However, total signal length mismatch is inherent when using connectors not specifically designed for broadside connection because the vias bringing the pairs to the surface will naturally be of different lengths. This subtle error component is evident in Figure 9.9. For simplicity, the dielectric separating the various layers is not illustrated.

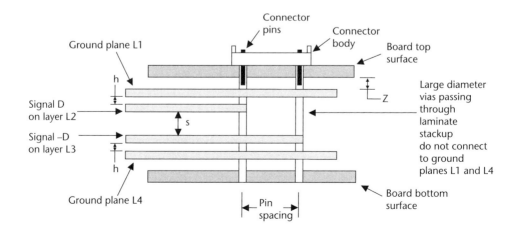

Figure 9.9 Broadside diff-pair length skew in pin field.

The signal D is seen to travel a distance equal to $(h + z)$ to reach the board's surface, while the $-D$ signal requires that distance plus an additional distance s.

Notice that even though in this example the vias pass completely through the PWB, the resulting length mismatch is s, the separation in height between the pairs. However, assuming identical connectors are used at both ends of the line and that they are on the same side of the board (such as in a backplane configuration), the total length mismatch is actually $2s$, as can be seen in Figure 9.10.

The $2s$ length difference can be a significant source of timing skew within a differential pair, especially if FR4 or other relatively high ε_r laminates are used. The use of such laminates forces s to be larger to achieve the desired differential impedance, and because [from (9.5)] the delay per inch will be higher with such laminates anyway, the result is a larger time skew than if a lower ε_r laminate were used.

For example, referring to Figure 9.9, assuming half-ounce, 8-mil-wide trace on FR4 ($\varepsilon_r = 4.5$), setting $h = 8.35$ mils and $s = 25$ mils yields $Z_{11} = 56.0\Omega$ and $Z_{12} = 6.12\Omega$. From Chapter 8, the system is somewhat loosely coupled, and from (8.14) $Z_{diff} = 2(Z_{11} - Z_{12}) = 99.8$. From (9.5), the timing skew is $t_d = 84.7\,\text{ps/in}\sqrt{4.5} \times 2(25\,\text{mils}) = 9$ ps.

This skew is acceptable at low data rates, but it becomes increasingly less tolerable at higher rates. For example, 9 ps of diff-pair skew represents just under 6.5% of the total worst-case eye opening when signaling at 2.5 Gbps (see [23]). This increases to about 8% of the opening when signaling at 3.125 Gbps (for example, see [24]).

Using a more advanced laminate having $\varepsilon_r = 3.5$ shrinks h to 6 mils and s to 9 mils, reducing the timing skew by better than 60% to 2.9 ps.

The $h = 8.35$ and $s = 25$ mils configuration is not the only 8-mil-wide FR4 arraignment that achieves 100-Ω differential impedance. Increasing s to 50 mils and reducing h to 6.65 mils to compensate results in $Z_{11} = 51.7\Omega$, $Z_{12} = 1.7\Omega$, again making $Z_{diff} = 100\Omega$. Doing this increases the $2s$ timing skew to 18 ps, but has the advantage of not requiring particularly tight alignment between the diff-pair layers L2 and L3. Additionally, as evidenced by Z_{12} being so small, this design is so loosely coupled that the dimension s is likewise noncritical. These things together mean this arraignment is easy to manufacture and will yield well, lowering cost. However, one penalty is the after-mentioned higher timing skew between the wires forming the diff pair.

In contrast, because the coupling is tighter in the $h = 8.35$, $s = 25$ mils stackup, the L2/L3 alignment and the dimensions s and h are somewhat more critical than

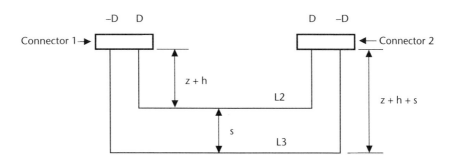

Figure 9.10 Broadside coupled differential pair skew.

the loosely coupled stackup, and this design would not be as straightforward to manufacture as one having larger separation. Misalignment of L2 relative to L3 lowers the coupling capacitance, resulting in higher impedance. Over etching one layer relative to the other has the same effect. It's worth noting that as a practical matter, this particular design is not overly critical and should have good manufacturing yields.

An inherent characteristic of broadside pairs is that the reference plane coupling is asymmetrical to the signal pairs. That is, L2 is more tightly coupled to L1 than it is to L4, and L3 is more tightly coupled to L4 than to L1. This means that more of the noise voltage present on L1 will be coupled onto the L2 signal than onto the L3 signal (and analogously for L4 coupling preferentially onto L3 but not L2). The result is unequal noise voltages on L2 and L3 unless L1 and L4 are very tightly connected. Tight L1/L4 coupling puts the same noise on L2 and L3, which will be rejected by an input receiver differential amplifier (as described in Chapter 8). From an implementation standpoint, this means vias must be placed at frequent intervals between L1 and L4 to ensure the noise voltage appearing on them has identical amplitude and phase.

9.5.2 Edge-Coupled Stripline

Edge-coupled stripline is probably the most familiar differential pair topology and was used in Chapter 8 to discuss coupling.

Side views showing two ways to form a 100-Ω differential pair are shown to scale in Figure 9.11. The traces are 5-mil-wide, half-ounce copper on FR4 ($\varepsilon_r = 4.15$). The topology on the left spaces the traces 5 mils apart edge to edge (called *5/5*); the topology on the right (*5/15*) spaces them 15 mils apart.

Just as was the case with broadside-coupled pairs, the degree of coupling between edge-coupled pairs is a factor in determining the overall stackup thickness for a given impedance. In this example, the 5/5 stackup is seen to be nearly twice as thick as the 5/15 stackup, and the coupling between traces is nearly 30 times as high.

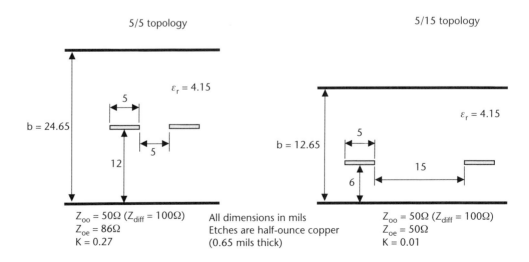

Figure 9.11 Two topologies to form a 100-Ω differential impedance pair.

The tight coupling forces the plate separation to be large to compensate for the mutual capacitance occurring between the traces. In the 5/5 topology, this results in the desired 50-Ω odd-mode impedance (100-Ω differential impedance), but also a relatively high 86-Ω even-mode impedance. This contrasts with the 5/15 topology, where the coupling is so low that the even- and odd-mode impedances are nearly identical at 50Ω.

This difference in even-mode impedances is significant because unless the differential signals are precisely $180°$ out of phase, the relatively high even-mode impedance will give rise to an unintended common-mode voltage. By having similar even- and odd-mode impedances, the 5/15 topology is more immune to this effect.

Taking this to an extreme suggests that s could be made arbitrarily large because doing so will have no further affect on Z_{oo} and Z_{oe}. Although in itself this is true, too large a separation raises the risk of different noise voltages being coupled onto the pair from the return planes. This occurs because the noise voltage present on each of the return planes varies by location but can be assumed to be identical (or nearly so) locally. Therefore, closely spaced traces will be exposed to the same noise, but widely spaced traces most likely will not. Keeping s small will help ensure the plane couples the same local noise to both traces. The noise then becomes common mode and will be rejected by the receiver.

Conductor loss is another factor that distinguishes the 5/5 and 5/15 topologies. For the dimensions shown, the 5/5 topology will have slightly lower loop resistance than the 5/15 topology, making the 5/5 conductor loss lower than that in the 5/15 topology. This comes about because, as described in Chapter 2 (for example, see Figure 2.8), the return path currents tend to spread out as the height h above the return plane increases. This means for a given frequency, the return currents will spread out more in the 5/5 topology than in the 5/15 topology, making the total loop resistance smaller. Analysis shows that at 1.25 GHz, the total conductor loss (trace plus return path) is about 3.5% lower in the 5/5 case. However, at this frequency on FR4 the difference in total loss (trace + return path + dielectric) will only be about half that, assuming the dielectric has a loss tangent of 0.02.

9.5.3 Edge-Coupled Microstrip

Besides being sandwiched between two plates and buried within the dielectric, differential pairs can also be created on the PWB's surface, in a microstrip configuration. This is shown in Figure 9.1.

Chapter 8 showed how the even- and odd-mode switching behavior of a signal pair affected the impedance and time of flight for microstrip traces. Differential pairs signal in the odd mode, suggesting that the even-mode characteristics of microstrips can be ignored.

But, as is the case with stripline, any imperfections in the launched waveform that makes the signals less than exact mirror images will excite the microstrip's even-mode behavior, resulting in degraded differential signaling. And again, just like with stripline, it's possible to obtain a desired microstrip differential impedance by adjusting the trace's height above the return plane (h) and its edge-to-edge spacing (s). Various combinations of s and h will result in the same differential impedance, but altering these in microstrip affects both the impedance and time of flight. This contrasts with stripline, where only the impedance is affected.

Intuitively, loosely coupled traces should have the least variation between even and odd modes, with the variation growing as s is made smaller. The first two entries in Table 9.3 confirm this.

The table's first entry shows a pair of 5-mil-wide, half-ounce microstrips on FR4 ($\varepsilon_r = 4.25$) covered in solder mask 0.7 mils thick over the copper ($\varepsilon_r = 3.2$) and spaced 25 mils apart edge to edge. The odd-mode impedance is 50Ω, resulting in a 100-Ω differential impedance. The coupling is loose enough to allow the odd and even impedances to be nearly identical. As a result of such loose coupling, the odd- and even-mode times differ by under 1%.

Reducing the spacing to 5 mils edge to edge as shown in the second entry has increased coupling, thereby lowering Z_{oo} and resulting in an odd- and even-mode timing difference approaching 3%.

In this example, the even-mode components will travel 8 ps/in faster than the odd-mode ones. This shows that unlike stripline, an even-mode signal will experience both an impedance and a temporal change relative to an odd-mode signal.

The third table entry shows that the height h must be increased to 4.2 mils to achieve a 100-Ω differential impedance with $s = 5$ mils. Doing so slightly reduces the flight time from the $h = 3.1$ mils case, but the odd-to-even-mode times vary by about the same percentage.

To avoid undesirable even-mode behavior, microstrips should be loosely coupled when used to form a differential pair. But again, similar to stripline differential pairs, if the microstrip spacing is too large, return plane noise will not equally couple to both lines forming the pair.

9.6   Summary

A stripline's time of flight is determined by the dielectric's ε_r, and not by the trace's topology (width, height above the return plane, or thickness). This contrasts with microstrip, where a trace's physical characteristics (including the spacing to adjacent traces) interplay to determine the time of flight.

Microstrip trace will be faster than stripline, as ε_{r_eff} will always be less than ε_r.

PWB fabricators can control Z_o in production to within 5% to 10% of a target specification by biasing the trace width, plating up the trace thickness, and adjusting the laminate and prepreg thickness.

Unless the user specifically requests otherwise, a thin coating (typically less than1 mil thick) of an epoxy-based solder mask is applied to all portions of the board surface where solder is not wanted. Solder mask has a loss tangent similar to FR4, but a lower ε_r. This lowers a microstrip's impedance, increases its capacitive

Table 9.3   Microstrip Characteristics

h (mils)	s (mils)	Z_{oe} (Ω)	Z_{oo} (Ω)	t_{de} (pS/in)	t_{do} (pS/in)
3.1	25	51	50	158	161
3.1	5	56	45	155	163
4.2	5	68	50	152	161

coupling to other traces, increases the time of flight, and increases dielectric losses over those of an otherwise identical exposed microstrip.

A *surface finish* is applied to all exposed copper on the board's top and bottom sides. This surface finish is usually conductive and prevents oxidation of the traces. It also increases the microstrip's thickness, thereby increasing the coupling to neighboring traces (especially when solder mask is present, as that increases the capacitance between traces).

Differential pairs can be formed as broadside or edge-coupled striplines or as edge-coupled microstrips. Broadside pairs increase the per-layer wiring density and make it easy to length match the traces forming the pair, but the inherent difference in via lengths when connecting to surface components will introduce a skew. Additionally, a sufficient number of vias must be used to connect together the reference planes to ensure the noise present on them is coupled equally to each trace forming the pair.

The differential impedance of a loosely coupled microstrip or stripline trace pair will change less with switching activity than a tightly coupled pair. This also applies to a microstrip's time of flight. Loose coupling is achieved by increasing the spacing *s* between traces, but care must be taken not to make the separation so large that the traces forming the pair are exposed to different noise.

References

[1] Miner, G. F., *Lines and Electromagnetic Fields for Engineers*, Oxford: Oxford University Press, 1996, p. 656.

[2] Sibley, M., *Introduction to Electromagnetism*, London: Arnold Press, 1996, p. 128.

[3] Cohn, S. B., "Characteristic Impedance of the Shielded-Strip Transmission Line," *IEEE Microwave Theory and Techniques*, Vol. MTT-2, No. 2, July 1954, pp. 52–57.

[4] Cohn, S. B., "Problems in Strip Transmission Lines," IEEE Microwave Theory and Techniques, Vol. MTT-3, No. 2, March 1955, pp. 119–126.

[5] Liao, S. Y., *Engineering Applications of Electromagnetic Theory*, New York: West Publishing Company, 1988.

[6] Djordjevic, A. R, et al., *LINPAR for Windows*, Norwood, MA: Artech House Publishers, 1999.

[7] Stanley, W., and R. F. Harrington, *Lines and Fields in Electronic Technology*, Englewood Cliffs, NJ: Prentice Hall, 1995.

[8] Mathsoft Engineering & Education, Inc., Cambridge, MA.

[9] The Math Works, Inc., Natick, MA.

[10] Kaupp, H. R., "Characteristics of Microstrip Transmission Lines," *IEEE Transactions on Electronic Computers*, Vol. EC-16, No. 2, April 1967, pp. 185–193.

[11] Bogatin, E., "Design Rules for Microstrip Capacitance," *IEEE Transactions on Components, Hybrids, and Manufacturing Technology*, Vol. 11, No. 3, September 1988.

[12] Walker, C. S., *Capacitance, Inductance and Crosstalk Analysis*, Norwood, MA: Artech House, 1990.

[13] Lee C. A., and G. C., Dalman, *Microwave Devices, Circuits and Their Interaction*, New York: Wiley and Sons, 1994 .

[14] Schneider, M. V., "Microstrip Lines for Microwave Integrated Circuits," *Bell System Technical Journal*, Vol. 48, No. 5, May/June 1969, pp. 1421–1443.

[15] Callarotti, A., and A. Gallo, "On the Solution of a Microstripline with Two Dielectrics," *IEEE Trans. Microwave Theory and Techniques,* Vol. MTT-32, No. 4, April 1984 pp. 333–339.

[16] Darwish, A., et al., "Properties if the Embedded Transmission Line (ETL)—An Offset Stripline with Two Dielectrics," *IEEE Microwave and Guided Wave Letters,* Vol. 9, No. 6, June 1999, pp. 224–226.

[17] Darwish, A, et al., "Effective Dielectric Constant of the Embedded Transmission Line (ETL)—A Multilayer Stripline-Like Structure," IEEE MTT-S International, Anaheim CA, June 13–19, 1999, *Microwave Symposium Digest,* Vol. 3, pp. 1249–1252.

[18] Pucel, R. A, D. J. Masse', and C. P. Hartwig, "Losses in Microstrip,"*IEEE Trans. Microwave Theory and Techniques,* Vol. MTT-16, No. 6, June 1968, pp. 324–350.

[19] Pucel, R. A, D. J. Masse', and C. P. Hartwig, "Correction to Losses in Microstrip," *IEEE Trans. Microwave Theory and Techniques,* Vol. MTT-16, No. 12, December 1968, pp. 1064.

[20] Denlinger, E. J., "Losses of Microstrip Lines," *IEEE Trans. Microwave Theory and Techniques,* Vol. MTT-28, No. 6, June 1980, pp. 513–522.

[21] Ross, R. F. G., and M. J. Hows, "Simple Formulas for Microstrip Lines," *IEEE Electronics Letters,* Vol. 12, No. 16, August 5, 1976, p. 410.

[22] Assadourian, F., and E. Rimai, "Simplified Theory of Microstrip Transmission Systems," *Proceedings of the IRE,* Vol. 40, December 1952, pp. 1651–1657.

[23] Infiniband Trade Association, "Infiniband Architecture Specification, Vol. 2, Release 1.1," November 2002.

[24] Vitesse Semiconductor Corp., "VSC7226-01 Double-Speed Multi-Gigabit Interconnect Chip Data Sheet, Revision 2.6," October 24, 2001.

CHAPTER 10

Surface Mount Capacitors

10.1 Introduction

Capacitors are ubiquitous on high-speed circuit boards, but design engineers often do not fully understand their electrical characteristics. This chapter discusses the operational behavior of surface mount ceramic capacitors in Section 10.2, and tantalum capacitors in Section 10.3. Through-hole devices and other capacitor types (such as metal film, mica, porcelain, or aluminum electrolytic) are not discussed.

10.2 Ceramic Surface Mount Capacitors

Surface mount technology (SMT) ceramic capacitors are comprised of multiple dielectric/electrode stacks fused together in a single package. Capacitors formed in this way are called *multilayer ceramic chip* (MLCC) capacitors. The number, size, and shape of the stacks are responsible for determining the capacitor's transient behavior, as they determine the *equivalent series inductance* (ESL) and *equivalent series resistance* (ESR). The choice of dielectric determines the capacitor's response to external effects such as mechanical shock, voltage bias (including the frequency response of that bias), and temperature. The dielectric material also has aging and ferroelectric characteristics, and its thickness determines the breakdown voltage, thereby setting the *working voltage* (WV) for a given capacitor.

10.2.1 Dielectric Temperature Characteristics Classification

As described in Chapter 3, the capacitance between two plates increases as the dielectric constant ε_r increases. Manufactures use high ε_r dielectrics to get as much capacitance as they can into the smallest possible SMT package. These dielectrics are ceramic blends that the Electronics Industries Association (EIA) has grouped into classes based on their temperature characteristic.

EIA specification 198 [1] defines a three-character scheme to indicate the dielectric temperature characteristic. The Class I dielectric coding is shown in Table 10.1, while Table 10.2 shows the coding for Class II dielectrics.

Class I dielectrics are the most stable, but capacitors in this class are generally limited to values under 10 nF, as they use ceramic blends (such as titanium dioxide) having a stable, but lower, ε_r (typically below 150). In commercial work, the C0G dielectric (still sometimes referred to by the older *NP0* classification) is the most commonly used Class I dielectric.

Table 10.1 Class I Ceramic Capacitor Temperature Coefficient Codes

First Character (Tempco, PPM/°C)		Second Character (Tempco Multiplier)		Third Character (Tempco Tolerance)	
C	0.0	0	−1	G	±30 PPM/°C
B	0.3	1	−10	H	±60 PPM/°C
A	0.9	2	−100	J	±120 PPM/°C
M	1.0	3	−1,000	K	±250 PPM/°C
P	1.5	4	−10,000	L	±500 PPM/°C
R	2.2	5	+1	M	±100 PPM/°C
S	3.3	6	+10	N	±2,500 PPM/°C
T	4.7	7	+100		
U	7.5	8	+1,000		
		9	+10,000		

Table 10.2 Class II Ceramic Capacitor Temperature Coefficient Codes

First Character (Lowest Temperature Rating)		Second Character (Upper Temperature Rating)		Third Character (Maximum Capacitance Change)	
X	−55°C	2	+45°C	A	1.0%
Y	−30°C	4	+65°C	B	1.5%
Z	+10°C	5	+85°C	C	2.2%
		6	+105°C	D	3.3%
		7	+125°C	E	4.7%
		8	+150°C	F	7.5%
		9	+200°C	P	10.0%
				R	15%
				S	22.0%
				T	+22, −33%
				U	+22, −56%
				V	+22, −82%

Class II dielectrics such as barium titanate are less stable but have a much higher ε_r (referred to as Dk or simply K by capacitor manufactures). Class II dielectrics are separated into two groups: The "Mid High K" (yielding the most stable Class II capacitors with values reaching the low μF range) and the volumetrically efficient but less stable "High K" dielectrics. High K dielectrics produce capacitors in the tens of μF, especially in the larger body sizes, but these capacitors exhibit stronger voltage and temperature effects.

Class II dielectrics are ferroelectric materials and exhibit a piezoelectric behavior. Class I dielectrics do not.

Ferroelectric behavior in a dielectric is roughly analogous to ferromagnetic behavior in inductors, except in the dielectrics case it's the relationship between the electric field and dielectric polarization that exhibits hysteresis.

Piezoelectric behavior causes ferroelectric dielectrics to be mechanically deformed by the application of an electric field. The inverse is also true: mechanical stresses resulting from equipment vibration or a sudden impact can result in the generation of noise voltages that may adversely affect circuit operation. Class I dielectrics should be used in applications where mechanical shock or vibration is present. This is especially important in low-level analog circuits and in high-gain amplifier circuits.

Referring to Table 10.1, the C0G designation has a temperature coefficient of zero, a temperature coefficient multiplier of −1, and a worst-case temperature coefficient tolerance of ± 30 PPM. Using (10.1), such a capacitor will vary by ± 0.3% across a 100°C temperature span.

$$C_2 = C_1 + TC(T_2 - T_1) \qquad (10.1)$$

where C_1 and C_2 are the respective capacitance values at temperatures T_1 and T_2. TC is the temperature coefficient with units of 10^{-6} for parts per million (PPM) and 10^{-9} for parts per billion (PPB).

The temperature characteristic for the High K Class II Z5U–type capacitor is found from Table 10.2 as having a lowest permissible operating range of +10°C, a maximum upper operating range temperature of +85°C, and a maximum variation in capacitance across that temperature range of +22% (at +10°C) to −56% (at +85°C), measured at zero dc bias. The capacitance specified on the data sheet is given at 25°C and for a low voltage bias, and it will be accurate to within the specified tolerance (e.g., ± 5%) at that temperature. In comparison, the Mid High K X5R–type capacitor varies by ± 15% across the same temperature span.

Note that the temperature variation in Class II dielectrics is highly nonlinear and thus will differ between manufacturers because each uses a unique blend of materials. The EIA 198 specification requires only that the capacitor characteristics fit within the overall window described in Table 10.2. It does not specify the behavior within that window.

Many capacitor manufacturers now offer Web-based or downloadable software that predicts the behavior of their MLCC capacitors over various environmental conditions. Figure 10.1 shows the performance across temperature of three dielectric blends from various manufactures, as reported by the manufacturers' software. For clarity, only two curves are shown for each capacitor type, representing the minimum and maximum span across the various manufacturers.

10.2.2   Body Size Coding

A four-digit scheme is used to describe the MLCC capacitor SMT package body size. The first two numbers indicate the capacitor's nominal length (defined as the electrode-to-electrode spacing), and the final two numbers describe its width. Both dimensions are in tens of mils (1 mil = 0.001 in = 0.025 mm). For example, a 0805 SMT capacitor is nominally 80 mils long (end to end) and 50 mils wide (2 mm × 1.25 mm).

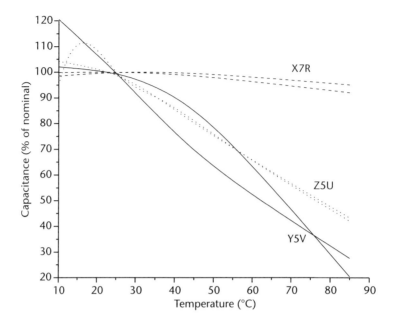

Figure 10.1   MLCC capacitor capacitance versus temperature.

10.2.3   Frequency Response

The impedance versus frequency behavior of a 10-nF MLCC capacitor is generalized in Figure 10.2. The broken curve shows the expected performance for a perfect capacitor: the impedance decreases by 10 times for each decade increase in

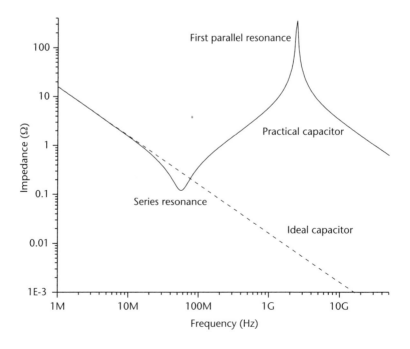

Figure 10.2   10-nF MLCC capacitor bode plot.

frequency. However, at high frequencies real capacitors exhibit the multiple resonances shown by the solid curve.

The capacitor's physical construction (especially the dimensions of the capacitor's plates) determines the resonances, but they are influenced by the placement of return paths on the circuit board. An electrical model incorporating RLC elements can be used to mimic the capacitor's frequency domain behavior.

Figure 10.3 shows the general electrical model for an MLCC capacitor. Resistances R_m and R_{df}, capacitor C_s, and inductor ESL form the series-resonant circuit responsible for the resonance at 55 MHz in Figure 10.2. At higher frequencies, those elements form a parallel-resonate circuit with capacitor C_p and resistor R_m. As shown, the first parallel resonance occurs at 2.5 GHz, but capacitors may have additional parallel resonances at higher frequencies.

Capacitor manufactures specify the series-resonate frequency either directly or by providing values for ESL, C_s, and indirectly for the sum of resistances R_m and R_{df}. Information on parallel resonance is usually not provided.

The series-resonate frequency is often of most concern in power-supply decoupling applications, but harmonics occurring near the first parallel-resonate frequency may also be of interest because currents appearing at those frequencies will not be shunted as effectively as those appearing near the series-resonate frequency. Referring to Figure 10.2, the capacitor will have an impedance of less than 1Ω for frequencies between 20 MHz and 200 MHz, but will have impedance about 10 times that at frequencies of 1 GHz and above. This situation also occurs when using MLCC capacitors to series couple (*ac couple*) high-speed signals into a differential receiver. A capacitor with low impedance at the fundamental frequency may still not perform well if it has a parallel resonance low enough to reduce important upper harmonics of the incoming signal.

Capacitor C_p is a modeling convenience to get circuit simulators to mimic the first parallel resonance seen in actual capacitors. It represents a portion of the self capacitance and a trivial parasitic shunt capacitance (generally under 5 fF) between the two surface mount pads and vias on the circuit board. Most manufactures specify neither C_p nor the first parallel resonance frequency. Instead, C_p is set to zero and all of the capacitance is lumped into C_s, as indicated by Figure 10.4.

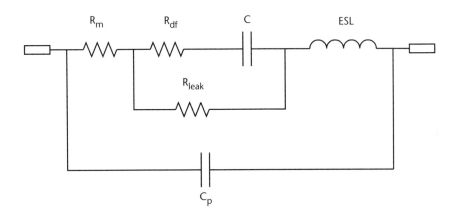

Figure 10.3   MLCC capacitor circuit model.

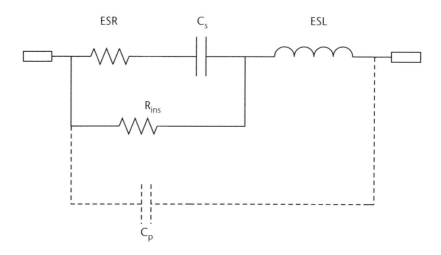

Figure 10.4   Reduced MLCC capacitor electrical model.

Orientation of the plates within the capacitor relative to the circuit board return path will influence the capacitor's serial and parallel resonant frequencies. Adding a return path directly underneath the capacitor body can beneficially raise the resonant frequency [2]. It's possible to reduce C_p by mounting the capacitor such that the plates are orthogonal to the circuit board return path, thereby moving the first parallel resonance to a higher frequency [3].

10.2.4   Inductive Effects: ESL

The plates forming the internal electrodes of the MLCC capacitor exhibit inductance, known as the equivalent series inductance (ESL). The ESL increases slightly with frequency but varies a great deal with package size and construction.

It was shown in Chapter 4 that inductance of a conductor increases with length and decreases with width. This general axiom applies to MLCC capacitors: assuming the same number of layers, a wider body style will have a lower inductance than a narrower one, especially if the wide capacitor has multiple electrodes [4].

A range of typical ESL values appears in Figure 10.5. These inductances do not include the additional inductance of the interconnecting vias on the circuit board and should be considered only approximate. The data comes from measurement and manufacturers' data and tactilely assumes the absence of a nearby return path. In practice, the inductance of these interconnecting vias adds to the ESL, causing the capacitor to resonate at a lower frequency.

Capacitors of the same value, voltage rating, and body size but made from different dielectric blends (such as X7R and Y5V) will exhibit different inductances. These differences are caused by the variation in the number of plates and the separation distance between them rather than by any innate inductive characteristics of the dielectric.

As Figure 10.5 illustrates, the smaller packages generally have lower ESL, but the differences can be surprisingly small, especially for higher capacitance values. For example, a 1-nF X7R 0805 has nearly the same ESL as an 0603, and the 1-nF,

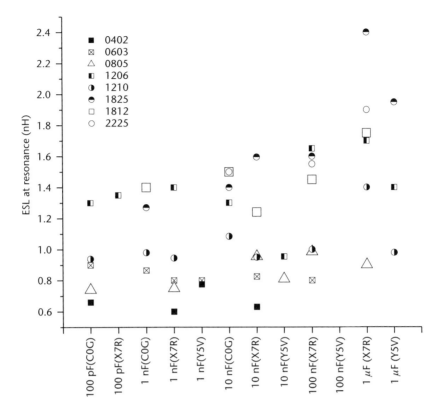

Figure 10.5   Typical ESL for various body styles.

10-nF, and 100-nF 0603 capacitors have essentially the same ESL. The 1-nF Y5V 0402 and 0603 capacitors have identical ESL. This comes about because these capacitors have the same length-to-width ratio, and as shown in Chapter 4 inductance is proportional to this ratio.

Also apparent is the increase in ESL as MLCC internal layer count increases to obtain high capacitance with lower K dielectrics. The 10-nF C0G capacitor in a 1210 package has a higher ESL than capacitors in the same package but of much higher values (such as the 1-μF Y5V), and the 1-μF X7R has significantly higher inductance than a 1-μF Y5V in an 1825 package.

10.2.5   Dielectric and Conductor Losses: ESR

Ohmic losses occur in the metal parts forming the capacitor, and in Class-II ceramics polarization loss of the dielectric also contributes to the overall loss. As presented in (10.2) and shown in Figure 10.4, these two separate loss mechanisms are summed into a single resistance called the equivalent series resistance (ESR):

$$ESR = R_{df} + R_{m} \tag{10.2}$$

where R_{df} represents the dielectric loss, and R_{m} represents the metalization resistance. The resistance of each varies with frequency, thereby making ESR change with frequency.

Polarization losses are insignificant in Class I dielectrics, making $R_{df} \approx 0$, but at low frequencies polarization loss is the dominant loss mechanism in Class-II dielectrics. Ohmic (resistive) losses in the metal electrodes and plates forming the capacitor are present in both Class I and Class II capacitors and at higher frequencies become the dominant loss mechanism.

Manufactures do not specify the metalization resistance, but do indicate the dielectric loss in terms of a *dissipation factor* (Df). Dissipation factor is identical to the loss tangent described in Chapter 3 but is usually given as a percentage. For Class I dielectrics, *Df* is usually taken as zero, but in Class II ceramics *Df* typically runs in the 1% to 4% range at low frequencies. At higher frequencies, it can approach 20% at in the High-K dielectrics.

At low frequencies (roughly the sub 10 KHz–100 KHz range), the *Df* of Class II dielectrics is initially flat or shows a slight decrease with increasing frequency but climbs at higher frequencies. At high frequencies (well over 1 GHz), some High-K dielectrics show a brief reduction in *Dk* after rising steadily, but *Dk* begins to climb again at even higher frequencies. The High-K dielectrics exhibit the highest *Df*, show the greatest variation with frequency, and show the onset of increasing *Df* at a lower frequency.

The relationship between *Df* and R_{df} is given by (10.3) [5], where C_s is the capacitance, *f* the frequency in hertz, and *Df* the dissipation factor.

$$R_{df} = \frac{Df}{2\pi f C_s} \qquad (10.3)$$

From (10.3) R_{df} is seen to fall with increasing frequency and will be small for frequencies in the megahertz range and above. For example, a 100-nF capacitor having a constant 2% Df will have an R_{df} of 32 mΩ at 1 MHz, and 320 μΩ at 100 MHz.

At moderate frequencies (typically 10 MHz to 30 MHz, depending on capacitor construction [5, 6]), skin effect (described in Chapter 2) causes the metalization resistance R_m to increase by $\sqrt{f}$ over its low frequency value. This increase occurs as R_{df} is falling toward zero, and at a high enough frequency R_m eventually exceeds R_{df}. ESR then climbs with a slope determined by the relative weights of R_{df} and R_m. For low-loss dielectrics, the ESR slope will increase nearly as the square root of frequency, but higher loss dielectrics will have a more gradual slope. This complex behavior is difficult to capture with a single datasheet parameter, so it's best to rely on measurements or manufacturers' software to obtain the ESR behavior for a given capacitor.

Figure 10.6 tabulates the typical frequency behavior of ESR for C0G, X7R, and Y5V capacitors. As shown, for a given capacitance, ESR is lower in the more stable dielectrics (C0G versus X7R and X7R versus Y5V), and ESR is lower in the higher capacitance values.

Intuitively, a capacitor having a wide and short body should have the lowest ESR because, as is the case with inductance, the electrode resistance increases with length and decreases with width. In general, this axiom holds true, but the choice of dielectric is a significant factor in determining the actual value. This is because from (10.2), R_m is only one of the factors in determining ESR. For example, according to one manufacturer's data, a 100-nF, X7R capacitor has an ESR of about 60 mΩ in an

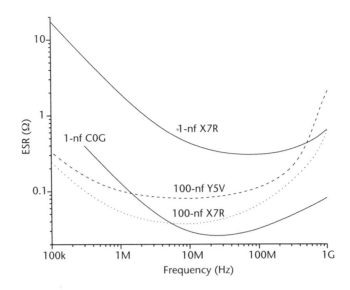

Figure 10.6   ESR versus frequency.

0805 package. Changing to a Y5V dielectric in the same package causes this to more than double to 150 mΩ. Capacitors in the 0805 and 1206 packages generally have similar ESR values (typically the 0805 is no more than 15% higher than the 1206), and the 0603 package has the highest value (typically 1.5 to two times the 0805 value). These are approximations, and there is a wide variation by dielectric type, capacitance value, and manufacturer.

Figure 10.7 illustrates the typical range at 25°C and 0-V bias of ESR, as reported by several manufacturers for three package types (0603, 0805, and 1206). Implicit in the figure is that the 0603 package has the highest ESR, and the 1206 has the lowest, with the 0805 laying between these limits.

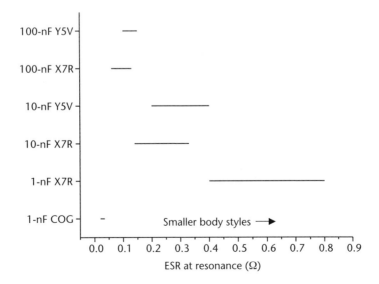

Figure 10.7   ESR for 0603, 0805, and 1206 body styles at 25°C, 0-V bias.

For example, a 10-nF X7R has ESR in the range from 140 mΩ (for a 1206 package) to 330 mΩ for the 0603 package. Changing to a Y5V dielectric increases this to 200 mΩ to 400 mΩ.

The general behavior of ESR as a function of temperature, bias voltage, WV rating, and package type is presented in Table 10.3.

For example, with an X7R dielectric, one can expect about a two times decrease in ESR as temperature increases from 25°C to 85°C and about a three times decrease when the bias voltage increases from 0V to 100% of the specified WV. Using a higher voltage capacitor (50 WV rather than 10 WV) reduces ESR by about 15%, and it's possible to obtain up to a further 15% reduction by using a wider package (e.g., moving from a 1206 to a 1210). These figures are generalizations across a very wide span of capacitances and body styles but illustrate the relative improvement possible. Table 10.3 can serve as a starting point when performing detailed analysis during the component selection process.

10.2.6   Leakage Currents: Insulation Resistance

An MLCC capacitor will conduct a leakage current when a dc bias is applied. A single parameter called the *insulation resistance* (IR), specified as an RC product with units of ohm-farads, is used to account for all of the leakage current. This leakage is due to dc conduction in the dielectric and increases as plate area increases. For this reason, higher valued capacitors have higher leakage currents. The insulating encapsulating material forming the capacitor body normally contributes trivial leakage. However, body leakage can become significant if improper cleaning leaves surface contamination on the circuit board or capacitor body, or if the capacitor absorbs moisture through cracks.

IR is not the resistor R_{leak} shown in Figure 10.5. Instead, R_{leak} is found by using (10.4).

$$R_{\text{leak}} = \frac{IR}{C_s} \tag{10.4}$$

For example, R_{leak} for a 100-nF capacitor is 10 GΩ if the IR is specified as 1 KΩ-F. As expected, R_{leak} is inversely proportional to capacitance, showing that leakage will increase as capacitance increases.

Table 10.3   Effect of Temperature, Bias, Package on ESR

Condition	X7R	Z5U	Y5V
Temperature ↑ from 25°C to 85°C	2X ↓	2.5X ↓	3X ↓
Bias voltage ↑ from 0 to 100% WV	3X ↓	15X ↓	20X ↓
WV ↑ from 16 WVdc to 50 WVdc (at 50% bias)	15% ↓	< 5% ↓	< 5% ↓
Package type (1206 → 1210)	< 15% ↓	< 10% ↓	< 20% ↓

For small values of capacitance, R_{leak} becomes too high to measure practically, so manufacturers usually specify an upper limit such as 10 or 100 GΩ. The value used for R_{leak} is either the value calculated by (10.4) or the upper limit specification, whichever is smaller.

Dielectric conduction increases with temperature, causing R_{leak} to fall, with the actual characteristics determined by the precise blend of the dielectric. High-K ceramics usually have lower resistivities and so show higher leakages.

Class-I dielectrics have the highest and least-changing IR specification and are the preferred ceramic capacitor choice for leakage-sensitive circuits such as sample and hold circuits, timing circuits, or some ac-coupling circuits.

10.2.7 Electrical Model

Most manufacturers use the simple series-resonate circuit shown in Figure 10.4 (without capacitor C_p) to model the high-frequency behavior of their capacitors. This model sums the resistive losses into a single resistor (ESR) and eliminates shunt elements R_{leak} and C_p. The absence of R_{leak} is inconsequential in most bypass and coupling applications, but not including capacitor C_p eliminates the often-onerous parallel resonant behavior described in Section 10.1.3.

The widely used model of Figure 10.4 is appropriate in many applications, but it can be made more suitable for high-frequency work by changing it into the model presented in Figure 10.3.

Capacitance C_p can be determined by measurement of the parallel-resonant frequency or occasionally obtained from manufacturers' software. As noted in Section 10.1.3, C_p is influenced by capacitor orientation and the proximity of circuit board return paths. Capacitance C_s is the nominal capacitance value minus C_p.

ESR can be determined from measurement or manufacturers' software for the frequency range of interest, remembering that Df is frequency dependent.

As shown in Figure 10.5, data sheet values for ESL are difficult to use properly because manufactures often don't describe the setup used to produce the inductance given in the data sheet. The presence of a nearby return path can lower ESL, thereby increasing the series-resonant frequency. User measurements that include return paths in the circuit board similar to that expected in the actual application are the best way to determine ESL. Nonetheless, Figure 10.5 is a good estimate and starting place.

Properly measuring ESR and ELS is not easy at frequencies in the hundreds of megahertz and beyond. Use of a vector network analyzer with special fixturing and a test method explicitly designed to remove fixturing errors is necessary [7, 8].

Finally, although not part of the capacitor model itself, the capacitance and inductance of the mounting pad and vias should be included as parasitic shunt elements for a proper high-frequency model.

Note that such a model should be used with care when performing time-domain simulations (such as the .TRAN simulation in SPICE) because this type of simulation tacitly assumes that the component values remain fixed with frequency. A frequency-domain simulation (such as the .AC simulation in SPICE) is more appropriate for determining high-frequency behavior, especially if the frequency-dependent elements are made to vary in the model with frequency in the simulation.

10.2.8   MLCC Capacitor Aging

The ferroelectric dielectrics used to form Class II capacitors show a logarithmic reduction in dielectric constant (and thus of capacitance) and an improvement in *Df* (loss tangent) over time. The precise dielectric blend determines the severity of the aging process and so will vary between manufacturers. The *Df* generally falls (improves) at a significantly faster rate then does the capacitance. Capacity and loss tangent aging is much more pronounced with the higher K dielectrics.

Class I dielectrics (such as C0G formulations) are much more stable and do not appreciably age.

The aging process in Class II dielectrics starts once the capacitor has cooled below its Curie point and is a result of the gradual realignment of the dielectric's crystal structure over time [9, 10]. A capacitor does not need to be in operation to age: a MLCC stored in a stockroom will see a reduction in capacitance over time, even in the absence of a bias voltage. A high temperature *bake out* (120°C to 150°C for up to several hours, depending on the manufacture and dielectric) can be used to restore the original manufactured capacitance value and reset the aging clock.

Manufacturers specify the decrease in capacity over time by quoting a *percent loss per decade hour* figure.

The last column in Table 10.4 shows the range of aging rates as specified by several manufacturers for various formulations. Quite a variation is evident. For example, depending on the manufacturer, a Y5V MLCC capacitor will age from between 3% to 7% per decade hour. After 100 hours, such a capacitor will have experienced two decades of time worth of loss: depending on manufacturer, a loss in capacity of between 6% and 14%. However, the more stable X7R ages at a rate of between 1% to 2.5% per decade hour and thus will have only lost between 2% and 5% of its capacity after 100 hours.

The logarithmic nature of the aging process means that MLCC capacitors show the most dramatic loss in capacity within the first 1,000 hours or so (about six weeks) after manufacture. Capacitor manufacturers take advantage if this to preage the capacitors to ensure that shipped units stay within tolerance for a reasonable time once at a customer's site. Indeed, IEC-384-9 requires that the capacitor stay within tolerance for at least 1,000 hours after it has cooled below the Curie point [11].

For example, after the first month of life, the Y5V capacitor must age for another decade (10,000 hours, nearly 14 months) before its value is reduced by another 3% to 7%. At that point, the capacity will be reduced from its original manufactured value by a total of between 9% and 21% (depending on the selected

Table 10.4   Characteristics of MLCC Capacitor Dielectrics

Type	Operating Temperature Range	Capacitance Tolerance	Capacitance Range	Aging
C0G	−55°C to +125°C	±10%	0.5 pF–10 nF	Nil
Z5U	+10°C to +85°C	+22, −56%	10 nF–1 μF	5%–7%
Y5V	−30°C to +85°C	+22, −82%	1 nF–22 μF	3%–7 %
X5R	−55°C+85°C	±15%	22 nF–10 μF	2%–3 %
X7R	−55°C to +85°C	±15%	100 pF–2.2 μF	1%–2.5%

manufacturer), but due to preaging by the manufacturer the customer receives a capacitor within the tolerance specification.

If heated high enough and long enough, the capacitance will rise and approach the original manufactured value of the capacitor. This may be higher than the data sheet value and actually exceed the high-side tolerance. Hand soldering with an iron or wave or reflow soldering can be enough to produce this unintentional bake out. Consequently, it's possible for MLCC capacitors mounted on freshly assembled and soldered PWBs, or boards that have been reworked, to have capacitors with higher than expected capacitance. And this capacitance will rapidly decrease with time because the aging clock may have been partially reset by the unintentional bake out.

10.2.9   Capacitance Change with DC Bias and Frequency

The dielectric in an MLCC capacitor experiences voltage stress proportional to the bias voltage placed across the capacitor and inversely proportional to the dielectric thickness. Because the dielectrics are thin, even a small voltage can expose the individual dielectric layers to very high electric fields. Manufacturers use the WV rating to specify the highest operating voltage a capacitor is allowed to experience, but dielectric voltage bias effects generally require the designer to operate an MLCC capacitor well below this value. Dielectric breakdown is typically tested at 2.5 times the rated WV for X7R, Y5V, and Z5U capacitors, but this varies with manufacturer, and operation above the WV is not permitted. In fact, the sum of the dc and ac (*ripple*) components of the waveform impressed across a capacitor must not exceed the dc WV limit, even momentarily.

Note that capacitors employed in low-frequency ac applications (below roughly 50 KHz) may need to significantly derate the WV [5].

Voltage bias has little effect on Class I capacitors, provided the voltage is below the rated WV of the capacitor. However, the dielectric constant and dissipation factor of Class II dielectrics decrease with applied dc voltage bias but increase when exposed to an ac bias [12]. The most significant changes occur in the High-K dielectrics, and, as frequency increases, these Class II ceramics experience a reduction in dielectric constant.

Capacitor vendors specify capacitance at low bias levels—often 0.5V or 1V—and not at the rated WV of the capacitor. As shown in Figure 10.8, exposing a Class II capacitor to voltages close to its rated WV will result in a capacitance reduction that can be quite significant, especially for High-K dielectrics (such as Y5V and Z5U). Figure 10.8 comes from data sheets and software of many manufacturers and shows the minimum/maximum performance range across those manufacturers for a 100-nF capacitor. Capacitors of other values will behave similarly. The dielectric's thickness and precise blend will determine the severity of the voltage bias effect and so varies by manufacture. One manufacturer may opt for a construction using fewer layers of a thinner dielectric, while another may use many thicker layers to achieve the same capacitance. For a given dielectric blend, the construction exposing the individual dielectric layers to the lowest electric field strength will be least affected by bias voltage. Figure 10.8 demonstrates that differences in construction result in significant variation in electrical operation between manufacturers of otherwise "identical" capacitors. For example, when exposed to a bias voltage equal to 50%

of the WV, a 100-nF Y5V capacitor will have between 25% and 70% of its zero bias (data sheet) capacitance, depending on the manufacturer.

10.2.10 MLCC Usage Guidelines

Choose C0G capacitors when capacitance stability is critical (such as PLL loop filters and in pulse timing circuits), and for best performance when ac coupling very high-speed signals (including ac termination of clock signals). These capacitors will cost about two times more then an X7R of the same capacitance value. C0G dielectrics do not require voltage or temperature derating and will not create piezoelectric noise when mechanically shocked. The frequency stability of their capacitance makes them excellent choices in ferrite or inductor PI filters.

Use X7R dielectrics in less critical timing and pulse-forming circuits, and when ac coupling moderate to high-speed signals and in critical power supply bypassing (such as PLL and oscillators). These capacitors are suited for use with ferrite PI filters, provided the filter's response is adjusted to account for the change in capacitance as a function of voltage, temperature, and especially frequency. As a general rule, operate an X7R at no more than 75% of its rated WV, and at that voltage expect a 5% to 10% reduction in capacitance at 50°C. These capacitors exhibit only small piezoelectric behavior and are suitable for use in the feedback loops of high-gain amplifiers.

Interchangeably, use Z5U or Y5V dielectrics in most power supply decoupling applications. Choose Y5V if its lower minimum temperature range is required. These capacitors cost roughly the same, but are about 25% less expensive then X7R dielectrics. Derate the WV by 50% or greater. Even when so derated, at 50°C expect

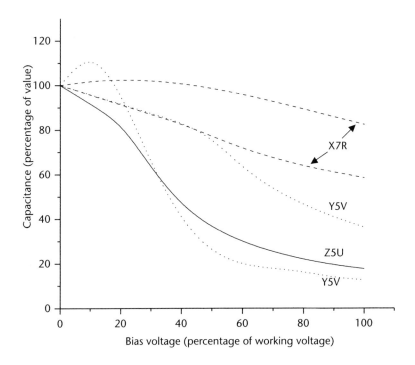

Figure 10.8 MLCC capacitor capacitance change versus percentage of WV.

the Z5U capacitor to have only 30% to 50% of the nominal datasheet specified capacitance. With the same derating, the Y5V dielectric can be expected to have capacitance in the 30% to 40% range. These capacitors show the most piezoelectric behavior and are not suited for coupling or bypass applications where mechanical stress or shock is present. They should not be used in high-gain feedback loops or as storage elements in sample/hold circuits, and they are not appropriate for use in ferrite PI filters unless their capacitance variability is swamped out by more stable capacitance (tantalum or possibly X7R ceramics).

Account for gradual loss in capacity due to aging, even in the absence of a bias voltage. This is especially important when using Z5U and Y5V dielectrics. C0G dielectrics do not appreciably age.

As a general rule, when operating at 75% of WV and at 50°C, expect an X7R capacitor to have 85% to 90% of its original capacity remaining after 10 years of life. When operating at 50% of its WV and at 50°C, expect Z5U and Y5V capacitors to have 30 to 40% of their original capacity remaining after 10 years of life.

Choose a larger body size (such as 0805 or 1206) to reduce ESR. To reduce ESL, select wider capacitors (such as 1825 or 2225) or capacitors that are wider then they are long (such as an 0508). These capacitors will also have lower ESR. Choose a body size as wide and as short as possible when using these capacitors in decoupling applications.

Smaller body sizes (0402 and 0201) save circuit board area but can be difficult to work with during circuit board assembly, rework, and debug. Because circuit board via inductance adds to ESL, these capacitors will not have appreciably lower loop inductances compared to some of the larger capacitors, unless high-end circuit board processing such as buried or blind vias is employed. When using standard circuit board processing, only select these capacitors to save circuit board space and not with the expectation of obtaining superior ESR or ESL. Also note that in standard circuit board processing, mounting pad and via restrictions will cause some of the area gain obtained by having a small body size to be lost.

10.3 SMT Tantalum Capacitors

Tantalum capacitors use a tantalum pentoxide film as the dielectric. Tantalum is a unique metal in that its oxide is electrically insulating and its thickness can be accurately controlled during manufacturing. Aluminum and niobium share these properties but tantalum has a higher dielectric constant than aluminum (26 versus 8.4), making tantalum capacitors volumetrically more efficient then aluminum electrolytics. Niobium has a dielectric constant about 1.6 times higher than tantalum, but higher leakage and lower breakdown voltage ratings have until recently prevented it from being a practical competitor to tantalum [13, 14].

10.3.1 Body Size Coding

Unlike ceramic capacitors, manufactures do not adhere to a uniform standard when specifying the physical dimensions of SMT tantalum capacitors. Most use a letter code ("A" through "D" being the most common) to signify the body size, but the actual dimensions vary between manufacturers. In addition to letter codes, some manufacturers signify body size with a four-digit code similar to that used in MLCC capacitors. For example, a manufacturer may interchangeably use "3216" and "Case size A." Unlike the MLCC capacitors, sizes of tantalum capacitors are usually specified in tenths of millimeters rather than thousandths of inches (*mils,* confusingly similar to *millimeters*). Thus, a 3216 tantalum capacitor is roughly 3.2 mm long by 1.6 mm wide (i.e., 126 mils by 63 mils).

The lack of body size uniformity means that the mounting pad land pattern recommended by each manufacturer may not be identical for capacitors with the same letter code case sizes. This complicates second sourcing, as capacitors with the same letter code body size coming from multiple manufacturers may not all attach to the PWB equally well.

10.3.2 Frequency Response

Compared to ceramics, tantalum capacitors have higher leakage and lower impedance at lower frequencies. It's apparent in Figure 10.9 that they also have a flatter resonance response: the resonance of the 4.7-μF tantalum capacitor is not as sharp as the 4.7-μF Y5V ceramic. This broad frequency response can be taken advantage of in power supply decoupling situations to prevent the power distribution system from appearing as a single, massive high Q capacitor at one or more discrete resonant frequencies.

In tantalum capacitors, capacitance does not appreciably change as frequency increases from dc to roughly 20 KHz or 30 KHz, but it falls off sharply at higher frequencies. In fact, the capacitance of SMT tantalum chip capacitors can be reduced over its dc value by more than 30% at 100 KHz and above [15].

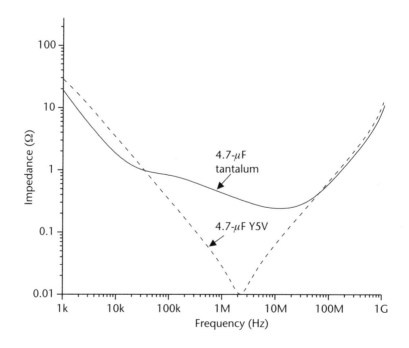

Figure 10.9 Impedance of tantalum and ceramic capacitor.

10.3.3   Electrical Model

The electrical model presented in Figures 10.3 and 10.4 for ceramic capacitors cannot produce the response for a tantalum capacitor shown in Figure 10.9. A distributed model, with the capacitance broken into three to five series chained RC segments, is better suited.

An appropriate model appears in Figure 10.10. The capacitance is usually evenly divided between each capacitor, but at least one manufacturer uses a binary weighting scheme [16].

10.3.4   Aging

Even under bias, the critical characteristics of tantalum capacitors (capacitance, Df, and leakage) do not change much over time. Along with the flatter resonance response described earlier, this stability in aging and when under bias is a major design difference when choosing between tantalum and high-value ceramic capacitors.

10.3.5   Effects of DC Bias, Temperature, and Relative Humidity

Tantalum dielectrics are more voltage and temperature stable than ceramics. Capacitance will generally increase only slightly (usually by less than 5%) when the capacitor is operated at the full WV rating and will increase slightly with increasing temperature. At 85°C the typical tantalum capacitor shows only a 5% to 10% increase in capacitance over the 25°C data sheet values.

For tantalum dielectrics, ESR decreases slightly with temperature: ESR at 85°C is about 10% to 20% less than the 25°C data sheet value. However, leakage strongly increases with temperature, and at 85°C ESR can be 10 times the leakage as at 25°C. Operating below the WV limit can offset this: operating at 40% of the WV can typically reduce leakage tenfold [17].

Increasing frequency causes ESR to fall, but ESR is essentially independent of bias voltage and is only slightly reduced as temperature increases.

Tantalum chip capacitors that are not hermetically sealed will see an increase in capacitance as relative humidity (RH) increases. Manufacturers often do not provide detailed capacitance versus RH data, but one manufacturer [17] indicates that capacitance can increase by as much as 12% as RH grows from 50% to 95% and will fall by 5% when RH shrinks from 50% to 25%.

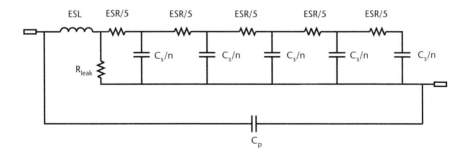

Figure 10.10   Tantalum electrical model.

10.3.6 Failure of Tantalum Capacitors

Service life of a tantalum capacitor is affected by three external conditions: operating voltage, operating temperature, and ESR-induced self heating. The capacitor's physical construction determines the severity of these effects and so varies greatly between manufacturers.

Tantalum capacitors usually fail by shorting rather than by opening. The short circuit can be catastrophic, especially if the tantalum is acting as bulk decoupling on a circuit board or within a power supply because, in these applications, a shorted capacitor ties the power supply rail to its return with a low-resistance connection. The resulting over current can cause damage to the power system or in extreme cases can cause charring of the circuit board near the shorted tantalum. Manufacturers have developed tantalum capacitors with built-in fuses to address these situations. However, such specialty capacitors generally have higher ESR than identical capacitors without fuses (two times is a good general rule), and there are fewer capacitance values available.

In general, operating life will improve exponentially when operating a tantalum capacitor below its WV and maximum temperature (usually specified as 85°C for industrial components).

Figure 10.11 illustrates the approximate improvement in operating life as a function of temperature and voltage as described by two manufacturers [18, 19]. Construction and material differences will cause capacitors from other manufactures to behave in their own way, but Figure 10.11 illustrates the typical improvement.

Curves for three temperatures are given, and by comparing against the 85°C curve the contribution due to voltage and temperature can be seen. For example, at 85°C reducing the operating voltage by 50% improves lifetime by eight times, but

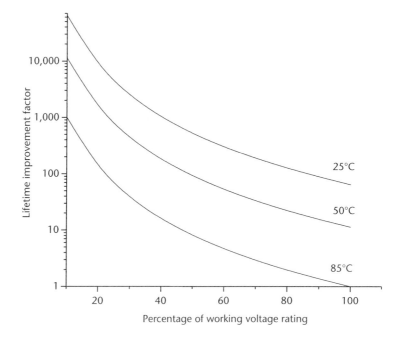

Figure 10.11   Approximate typical tantalum lifetime improvement with voltage and temperature.

operating the same capacitor at 50% of the WV and 50°C improves lifetime by a total of 90 times, an improvement of more than an order of magnitude over that obtained by just reducing the operating voltage. Manufacturers refer to operation at temperatures below 85°C and voltages below the rated WV as *derating* the capacitor.

10.3.7   ESR and Self Heating: Voltage and Temperature Derating

A capacitor's internal temperature increases as it dissipates power during charge/discharge cycles. The charge/discharge current (referred to by some manufacturers as *ripple current*) causes power to be dissipated in the dielectric's ESR, generating the heat. A prolonged exposure to enough heat will reduce the capacitor's life. The permissible power is set individually by each manufacture for each of their packages sizes, but in general, the larger case sizes such as "C" and "D" can dissipate more power and have lower ESR then the smaller "A" and "B" cases sizes.

The power dissipated in a capacitor's ESR is given by (10.5), which shows that power can be reduced either by limiting the ripple current I or by reducing the operating voltage E.

$$P = I^2 ESR = \frac{E^2}{ESR} \qquad (10.5)$$

Placing a resistance in series with the capacitor will lower ripple current. While this is possible in some timing circuits, it's hardly practical in decoupling applications because doing so greatly reduces the capacitor's ability to instantaneously dump charge in response to a large switching event. The alternative is to reduce the operating voltage, or, conversely, to select a capacitor with a higher WV. As was shown earlier, this will also improve lifetime.

In fact, manufacturers suggest that operating voltage be limited to no higher than 50% to 75% of the rated WV for normal capacitor operation and 30% for low-impedance circuits, such as power supply decoupling or pulse-type circuits that do not have a series resistor [17, 19]. For example, when decoupling a 3.3-V power rail, a tantalum with at least a 10-V WV should be selected. However, the higher voltage capacitors have higher leakage, so derating the WV too aggressively may result in unacceptably high leakage currents. This is usually not a problem for decoupling applications (although low-power applications need to exercise care), but the higher leakage can be significant in timing and pulse-type circuits.

10.3.8   Usage Guidelines

Tantalum capacitors are polarized, and the applied voltage must not be reversed. Doing so will cause the capacitor to become a low resistance, thereby conducting enough current to generate sufficient heat to permanently short (often with spectacular results). Capacitor manufacturers have unique guidelines as to the amount of reverse bias their capacitor can withstand. However, the safest policy is not to allow any reverse bias (including ac ripple riding on top of a dc bias) if interoperability between manufactures is desirable.

Operate the capacitor at reduced voltages. As a general rule in decoupling power supply applications, select the capacitor WV to be at least three times the power supply voltage. This will ensure interoperability between manufacturers. Don't select too high a WV if leakage is a concern (such as in low power or timing circuits). Derate the WV by more than three times to further improve lifetime.

If possible, operate the capacitor at reduced temperatures. The capacitor's internal temperature will rise above ambient due to self heating as large currents are switched. Orientating the capacitors so as to maximize airflow and using capacitors with large case sizes can mitigate the negative effects of high temperature and so improve lifetime.

Consider the use of capacitors with built-in fuses if the application cannot withstand a catastrophic failure of the capacitor resulting in a low-resistance short across the power rails. If the higher cost and ESR of these capacitors pose a problem, use resistors, FETs, fuses, or resetting circuit breakers in the power system to limit or interrupt current in the event of a tantalum failure.

Use the largest case size possible, remembering that cases with the same letter code will not necessarily be the same size from all manufacturers. This is especially true of the larger case sizes. Capacitors in larger cases have reduced ESR and better power dissipation.

Consider ultralow ESR capacitors in critical power supply decoupling applications. The selection of WVs and capacitances are not as great as with traditional tantalum capacitors, but the reduced ESR (often as much as two times lower than standard capacitors) can be helpful in situations when high current transients must be supplied quickly.

10.4   Replacing Tantalum with High-Valued Ceramic Capacitors

There is an obvious difference in the resonance behavior between the high-capacity ceramic and tantalum capacitors apparent in Figure 10.9. The strong resonance of the high-capacity ceramic capacitor contrasts sharply with the more gradual resonance of the tantalum. This suggests they will behave differently when used for PWB power supply decoupling.

Simulation results of a distributed section of power/ground planes presented in Section 6.6.2 and shown in Figure 6.20 appears next for a 350 mil by 350 mil power/ground plane section bulk decoupled with either a 4.7-μF tantalum or a 4.7-μF ceramic capacitor. The ceramic capacitor used the circuit was presented in Section 10.5, with an ESR of 0.02Ω and ESL = 1 nH. The tantalum model appearing in Figure 10.10 was used with ESR = 1.04Ω for each section and ESL = 1.8 nH. The shunt capacitance C_p was set to 5 pF for each capacitor.

The tantalum impedance is seen to be somewhat higher than the ceramic but has a significant series resonance occurring at about 80 MHz. The ceramic-series resonance occurs at just over 2 MHz (visible in Figure 10.9) and its first parallel resonance (causing an impedance increase) is visible in Figure 10.12 at approximately 84 MHz. However, as demonstrated in Figure 10.13, it's not possible to generalize the overall noise response by examining only this narrow band of frequencies.

The figure shows the noise voltage measured on a 350 mil by 350 mil isolated region on a larger circuit board and modeled as in Figure 10.12. This isolated region

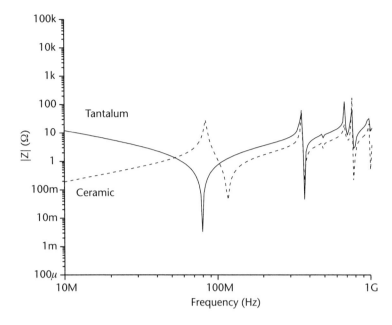

Figure 10.12 Impedance response of a small segment of power/ground planes with a 4.7-μF tantalum and ceramic capacitor.

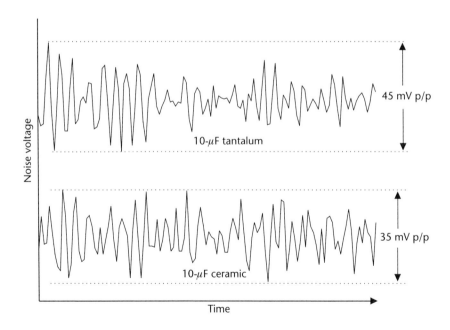

Figure 10.13 Measured noise voltage using the capacitors in Figure 10.12.

supplies 3.3V to power the PLL of an ASIC running at 156 MHz and is decoupled with several small-valued ceramic capacitors and one 4.7-μF capacitor. The power supply's response with a 4.7-μF tantalum capacitor is shown in the figure's top portion, while the figure's bottom part illustrates the response when the tantalum is replaced with a 4.7-μF ceramic. In this application, noise voltages appear well above the 80-MHz to 85-MHz resonance band, and the ceramic capacitors lower overall impedance at high frequencies results in less noise on the planes.

References

[1]   Electronics Industries Alliance "Ceramic Dielectric Capacitors, Class I, II, III, IV, Part I: Characteristics and Requirements," ANSI/EIA 198-1-E-97.

[2]   Lakshminarayanan, B., et al., "A Substrate-Dependent CAD Model for Ceramic Multilayer Capacitors," *IEEE Trans. On Microwave Theory and Techniques*, Vol. 48, No. 10, October 2000, pp. 1687–1693.

[3]   "SRF & PRF and Their Relation to RF Capacitor Applications," Technical Note, Johanson Technology, Inc., January 17, 1999.

[4]   Schaper, L., and G. Morcan, "High Frequency Characteristics of MCM Decoupling Capacitors," *Proc. IEEE 1996 Electronic Components and Technology Conference*, pp. 358–364.

[5]   "AC Load of Ceramic Multilayer Capacitors," Application Note, Philips Components, Document No. 9398-084-18011, June 2000.

[6]   "ESR Losses in Ceramic Capacitors," Application Note, American Technical Ceramics, Inc.

[7]   Green, H., "Characterizing Chip Capacitors at Ultra-High Frequencies," *Test & Measurement World*, June 1989, pp. 95–104.

[8]   Li, Y. L., et al., "A New Technique for High Frequency Characterization of Capacitors" *Proc. 48th Electronic Components & Technology Conference*, Seattle, WA, May 25–28, 1998, pp. 1384–1390.

[9]   "Understanding Chip Capacitors," Applications Booklet, Johanson Dielectrics, Inc., 1985.

[10] "Tantalum and Ceramic Surface Mount Capacitors," Applications Booklet, Kemet Corporation, document F-3102E 7/99, pp. 32–37.

[11] "Fixed Capacitors for Use in Electronic Equipment: Part 9," IEC-60384-9(1988-06) International Electrotechnical Commission.

[12] Galliath, A. P., "Technical Brochure," Novacap Corporation.

[13] "Introduction to Tantalum Capacitors," Vishay Sprague, Document No. 40035 21, July 2000.

[14] "Kemet News," Kemet Corporation, August 29, 2001.

[15] "Tantalum Capacitors," Applications Note, Thompson-CSE.

[16] "Kemet Spice Software," Version 2.0.1, Kemet Electronics Corp.

[17] "Tantalum and Ceramic Surface Mount Capacitors," Kemet Corporation, Document No. F-3102E 7/99, pp. 4–14.

[18] "Tantalum Capacitors" Applications Note, Hitachi AIC, Inc.

[19] "Notes on the Correct Use of Tantalum Capacitors," Applications Note, NEC Corporation, Document No. ECC0332EJ1V0UM00, 1999.

Appendix: Conversion Factors

To Get From	To	Multiply By
Inches	Mils	1,000
Inches	Meters	25.4×10^{-3}
Inches	Centimeters	2.54
Inches	Millimeters	25.4
Inches	Microns	$25.4 \times 10^{+3}$
Inches	Angstroms	$25.4 \times 10^{+6}$
Microns	Meters	1×10^{-6}
Angstroms	Meters	1×10^{-9}
Nepers	Decibels	8.686
Decibels	Voltage ratio	$10^{-\frac{dB}{20}}$
Voltage ratio	Decibels	$20\log\left(\dfrac{V_o}{V_i}\right)$
F°	C°	$\dfrac{F° - 32°}{1.8}$
C°	F°	$1.8°C + 32°$

About the Author

Stephen C. Thierauf is chief scientist (technology) at Signal Integrity Software, Inc. Previously, he served as senior consulting hardware engineer at Fabric Networks/ Infiniswitch Corporation, where he was responsible for the signal integrity design, analysis, and debugging of very high signal count/high-performance backplanes and line cards. While working at Compaq Computer and Digital Equipment Corporation, he was a senior member of the technical staff responsible for the circuit design and signal integrity analysis of integrated high-speed I/O circuitry, as well as circuit board and micropackage level interconnect on the ALPHA microprocessor. Formerly a visiting scholar at Northeastern University, he holds a B.S. in electrical engineering technology from Wentworth College of Technology, Boston, Massachusetts. He has coauthored five papers regarding the Alpha microprocessor and has contributed to the book *Design of High Performance Microprocessor Circuits* (IEEE Press).

Index

For further information on these and other Artech House titles, including previously considered out-of-print books now available through our In-Print-Forever® (IPF®) program, contact:

Artech House Publishers
685 Canton Street
Norwood, MA 02062
Phone: 781-769-9750
Fax: 781-769-6334
e-mail: artech@artechhouse.com

Artech House Books
46 Gillingham Street
London SW1V 1AH UK
Phone: +44 (0)20 7596 8750
Fax: +44 (0)20 7630 0166
e-mail: artech-uk@artechhouse.com

Find us on the World Wide Web at:
www.artechhouse.com